3-80
206 559

8011
Physics

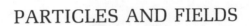

PARTICLES AND FIELDS

Readings from
**SCIENTIFIC
AMERICAN**

PARTICLES
AND FIELDS

With an Introduction by
William J. Kaufmann, III
Department of Physics
San Diego State University

W. H. Freeman and Company
San Francisco

Some of the Scientific American articles in *Particles and Fields* are available as separate Offprints. For a complete list of articles now available as Offprints, write to W. H. Freeman and Company, 660 Market Street, San Francisco, California 94104.

Library of Congress Cataloging in Publication Data

Main entry under title:

Particles and fields.

    Includes bibliographies and index.
    1.   Particles (Nuclear physics)—Addresses, essays, lectures.   2.   Quarks—Addresses, essays, lectures.
3.   Unified field theory—Addresses, essays, lectures.
I.   Kaufmann, William J.   II.   Scientific American.
QC793.28.P38   539.7′21   80–10669
ISBN 0-7167-1233-4
ISBN 0-7167-1234-2

Copyright © 1953, 1957, 1964, 1974, 1975, 1976, 1978, 1979, 1980 by Scientific American, Inc.

No part of this book may be reproduced by any mechanical, photographic, or electronic process, or in the form of a phonographic recording, nor may it be stored in a retrieval system, transmitted, or otherwise copied for public or private use, without written permission from the publisher.

Printed in the United States of America

9 8 7 6 5 4 3 2 1

PREFACE

Something very exciting is happening in physics. We are developing a new picture of matter. We are at the brink of a new concept of reality. We have begun to glimpse remarkable symmetries and profoundly beautiful properties of matter that were unimagined only a few years ago.

Revolutions of this magnitude are extremely rare. It seems to me that, throughout all recorded history, there are only three other scientific breakthroughs that rank with the discoveries of the 1970s. In the eighteenth century, Antoine Laurent Lavoisier argued that all substances are composed of *chemical elements*. In the nineteenth century, John Dalton spearheaded the notion that chemical elements are composed of *atoms*. Each chemical element has its own unique type of atom that can combine with other atoms to produce the wide variety of substances we find in the world around us. And finally, at the beginning of the twentieth century, Ernest Rutherford and Niels Bohr discovered that an atom consists of a dense *nucleus* orbited by tiny particles called electrons.

Each of these concepts constituted a fundamental insight into the nature of matter. They changed the way we think about the world around us. And each insight carried enormous powers that dramatically affected the course of humanity. The picture of reality that emerged from the concepts of elements, atoms, and nuclei forms the basis of all modern chemistry and physics. Our civilization—our political, social, and economic structures—all depend critically on the technology and industry that have blossomed from this chemistry and physics.

In the 1970s, it became clear that there are basically only two types of particles: *leptons* and *quarks*. Leptons come in various "flavors" (electron, muon, neutrino, etc.). And so do quarks (up, down, strange, charm, etc.). But quarks also have "color." By exchanging color, quarks are bound together to form all the kinds of particles that are associated with the nuclei of atoms (protons, neutrons, pions, etc.).

These notions sound confusing and fantastic at first glance. The terms "flavor" and "color" have nothing to do with our senses of taste and sight. Instead, they are part of the amusing terminology that has been coined to describe newly discovered properties of matter. Just as "atoms" and "nuclei" dominated physics in the first part of the twentieth century, surely "color" and "flavor" will play a very significant role in the years to come.

Shortly after the birth of atomic and nuclear physics, scientists were delighted to learn that they could finally understand why the sun shines. Answering the childlike question "Why does the sun shine?" requires some basic understanding of the nuclei of hydrogen and helium atoms. At the sun's center, hydrogen nuclei are fused together to form helium in a process that releases substantial amounts of energy.

When this thermonuclear process was first formulated in the 1920s, scientists were commonly asked if these reactions could ever be artificially reproduced here on earth. Was it possible that humanity might someday tap the enormous energies buried inside the nuclei of atoms? These musings were roundly ridiculed as idle fancies. Many world-renowned scientists dismissed these ideas as complete impossibilities forever banished to the dreamy realms of science fiction.

Two decades later, thousands of people were incinerated at Hiroshima and Nagasaki. Thermonuclear weaponry today dominates the balance of international power. And nuclear energy has become a heated topic of public debate.

As you will learn in the pages of this book, we have recently discovered that the strong nuclear force that binds protons and neutrons together in the nucleus is merely a faint shadow of the much more powerful "color force" between quarks inside these protons and neutrons. This is a totally unexpected development. Deep inside objects that were once called "elementary particles" there is a force more powerful than anything we have ever imagined.

Most physicists would probably scoff at the idea that humanity might someday tap the color force. But surely it is difficult to imagine how scientists of the 1920s could have envisioned the specter of today's worldwide arsenal of thermonuclear weapons.

Although we have no means of predicting the future, it is certain that today's developments in physics will play a decisive role in tomorrow's society. History is crystal-clear on this point. Since the time of Isaac Newton, nearly every major scientific discovery has had a direct effect on civilization.

The modern scientist has a profound moral obligation to bring scientific developments to the attention of the general public. This is especially true in a democracy, where the public is the electorate. Because of the pace of technological development, society can no longer afford the luxury of ignorance. Knowledge of the physical world bestows awesome powers. To understand the secrets of atoms and galaxies is to become like the gods. Whether we use this knowledge for the betterment of humanity or for the destruction of our planet is entirely a matter of our own free choice.

March 1980   William J. Kaufmann, III

CONTENTS

*Nobel Laureate in Physics

Note on cross-references: References to articles included in this book are noted by the title of the article and the page on which it begins; references to articles that are available as Offprints, but are not included here, are noted by the article's title and Offprint number; references to articles published by SCIENTIFIC AMERICAN, but which are not available as Offprints, are noted by the title of the article and the month and year of its publication.

PARTICLES AND FIELDS

INTRODUCTION

There are many things in the world. The multitude of objects that we encounter at every waking moment is absolutely staggering. And the manifold ways in which these objects behave and interact are equally overwhelming. One of the primary functions of the brain is to selectively sort out the myriad experiences that constantly bombard our senses. Modern psychology teaches us that in order to function in the world, to perform even the simplest tasks such as eating a sandwich or reading a book, we must focus our attention on certain sensory input while deliberately ignoring or habituating to a host of extraneous sensations. A person who fails to sort out sensory input or is unable to selectively filter the enormous mass of data assailing the brain is totally overwhelmed and is rendered catatonic.

To accomplish this feat of sensory processing, the brain must have the ability to form generalized concepts of both objects and processes. For example, the sane person recognizes certain spherical, reddish objects as apples whether they are hanging from a tree or displayed in a bin at a market.

We share this necessary level of abstraction with the animal world. A chimpanzee has no trouble identifying objects that we call bananas. But the human mind has the extraordinary ability of extending this intellectual activity to profoundly esoteric levels. This single ability, perhaps more than any other characteristic, is responsible for the unique position that humanity occupies among the living creatures on this planet.

From the moment of birth, our daily experiences strongly enforce the notion that reality is comprehensible. The fact that a rock released from your hand always falls down or that the moon goes through its phases every 29½ days implies order rather than chaos to the rational human mind. To discover this order, to understand the basic and underlying qualities of all physical objects, to comprehend the fundamental principles that dictate the behavior of reality: this is the business of science.

In many respects, physics is an intellectual activity that extends the process of consciousness to the most abstract and fundamental levels. The physicist wants to know what the world is made of and how it works. The history of this quest has been characterized by finer and finer subdivisions in an increasingly microscopic realm. We first discovered that the world is composed of 92 naturally occurring elements. All objects, regardless of their origin or individual properties, are constructed from these chemical elements.

We then learned that these elements are made of atoms, which are composed of particles called protons, neutrons, and electrons. The protons and neutrons together constitute the nucleus of an atom, while the electrons orbit this nucleus—like tiny planets about a submicroscopic sun.

In recent years, it has become increasingly clear that particles such as protons and neutrons are not fundamental entities. The preponderance of evidence indicates that protons and neutrons have an internal structure. The proton and neutron are composed of elementary objects called *quarks*. And apparently, this is where the process of submicroscopic subdivision stops. Quarks give every indication of being truly elementary particles. They behave like point masses with absolutely no internal structure.

Electrons (and closely related particles such as the muon and the neutrino) also appear to be fundamental entities. No matter how hard we probe with our highest-evergy accelerators, we find no indication of any internal structure.

Electrons and their relatives comprise one family of particles that we call *leptons*. Quarks, of which there are several types, comprise a second family of truly elementary particles. Everything in the universe is constructed from only these two types of particles: leptons and quarks. The century-old investigative process of finer and finer subdivision has come to an end. And at the conclusion of this intellectual journey, we find that these two families of particles exhibit profoundly beautiful symmetries. Explaining these properties, demonstrating the beauty and simplicity of what we have discovered, is the central goal of this book.

We begin with an entertaining and lively article by one of the greatest physicists of the twentieth century. As if to epitomize the central issue of our quest, Schrödinger chose the title "What Is Matter?" It is an excellent overview of some of the basic developments of the twentieth century. We clearly see the struggle with semantics because our language was developed in the macroscopic world. No wonder there are difficulties in using this same language to describe processes and properties of the submicroscopic world of atoms! And we forgive our ancestors' idiosyncrasies such as walking around with radioactive isotopes strapped to their wrists.

With the second article, we shift our attention from particles (what the world is made of) to fields (how the world works). Like the preceding article, Dyson's presentation is also somewhat dated. For example, the chart of fundamental particles should be regarded as an amusing historical anecdote. Nevertheless, I have chosen to include these older articles because they contain significant amounts of relevant background material. In addition, we clearly see how some of the greatest human minds began struggling with the perplexing issues that faced physicists in the 1950s and 1960s.

While reading Dyson's article, it is helpful to anticipate some ideas that occur explicitly later in the book. Foremost is the realization that there are four forces in nature: (1) gravitation, (2) electromagnetism, (3) weak nuclear force, and (4) strong nuclear force. The first two are long-range forces with which we have some familiarity in our everyday lives. The remaining two forces are extremely short-range; their effects are felt only over subatomic distances. The weak nuclear interaction occurs in certain types of radioactive decay, such as the decay of a neutron into a proton (which occurs with the emission of an electron and an antineutrino). The strong nuclear interaction is the force that binds protons and neutrons together inside nuclei. In later articles we shall come to realize that the weak interaction arises when quarks change "flavor," while the strong interaction is a result of quarks exchanging "color."

Another important issue to keep in mind is the central idea behind quantum field theory. When particles interact, they exchange energy and momentum. But quantum theory tells us that energy and momentum are quantized. They exist in small, discrete amounts or packets called *quanta*. Therefore, at the most fundamental level, we describe the interactions between particles by saying that they exchange other particles. This process is said to be "virtual" because we have no way of observing particles while they are being exchanged.

This idea of virtual exchange appears explicitly in the third article, by Gell-Mann and Rosenbaum. Today we find it very useful to draw pictures (called *Feynman diagrams* after their inventor, Richard P. Feynman) that epitomize

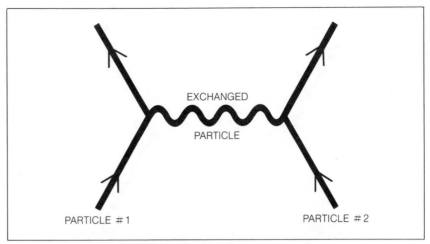

A FEYNMAN DIAGRAM is a useful way of visualizing a subnuclear event. Two particles interact by exchanging a third particle. Each force in nature involves the exchange of specific kinds of particles.

the interaction. An archetypal example is shown on this page. Each of the four forces in nature is mediated by particles that are exchanged in this fashion. For example, two particles exert a gravitational force on each other by exchanging gravitons. Similarly, two particles interact electromagnetically by exchanging photons. And in Yukawa's theory of the strong nuclear force, two nucleons (either protons or neutrons) interact by exchanging pions. Although we cannot observe these particles while they are being exchanged, they are nevertheless bonafide particles that can be manufactured and can exist in the real world.

In many respects, the fourth article, by Chew, Gell-Mann, and Rosenfeld, is a continuation and updating of the third. It is the last of the introductory selections. Reading between the lines, we see light at the end of the tunnel. The zoo of "elementary particles" has proliferated to such an extent that significant relationships and symmetries finally begin to appear. In fact, while the manuscript for this article was being prepared, one of the authors, Murray Gell-Mann, and George Zweig at Cal Tech were proposing that *all* strongly interacting particles are composed of quarks. The Dark Ages of particle physics was finally drawing to a close.

More than a decade separates the first four introductory articles from the remaining six, which focus entirely on the most modern developments. Although the quark hypothesis was obviously very attractive from the beginning, a number of troublesome details had to be worked out before prominent physicists could feel comfortable about presenting these ideas in the pages of *Scientific American.*

The fifth article, the first of these recent papers, is also the only article with any substantial discussion of experimental techniques. In assembling this anthology, I specifically avoided dealing extensively with hardware, such as accelerators, bubble chambers, storage rings, and so forth. The reader should not misinterpret my selectivity as a denigration of experimental work. Very much to the contrary, experiment and theory go hand in hand, as we see clearly demonstrated in Drell's article. Primarily because of space limitations, I simply chose to focus on theory, concepts, and ideas. Nevertheless, Drell gives a good overview of some of the crucial experimental work that currently goes on in this branch of physics.

My primary reason for choosing Drell's paper is that he skillfully and succinctly summarizes the major developments of the late 1960s and early 1970s. Although similar summaries appear in the next three articles, Drell's presentation is exceptionally lucid.

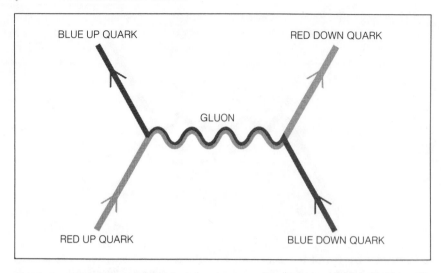

BLUE UP QUARK

RED DOWN QUARK

GLUON

RED UP QUARK

BLUE DOWN QUARK

QUARKS ARE BOUND TOGETHER by exchanging gluons. In this example, a red up quark interacts with a blue down quark by exchanging a red-blue gluon. Interactions of this type underlie the *color force* that keeps quarks permanently confined inside baryons and mesons.

The important concept of quark "color" is introduced by Drell. This use of the term has nothing to do with eyes or vision. *Color* in this context is simply the name given to one of the attributes, or *quantum numbers*, possessed by quarks. This quantum number can assume one of three values. There is, however, no general consensus on the names of these values. Some authors in this book (Drell, Glashow, and Johnson) use the subtractive primaries: red, yellow, and blue. Other people (Nambu, Gell-Mann, and I) prefer the additive primaries: red, green, and blue. In going from one article to the next, the reader should try to avoid being confused by an author's choice of color names. And, of course, antiquarks have anticolors, such as antired, antigreen, and antiblue.

The sixth article is, in some ways, a continuation of the fifth. Glashow, one of the leading theoreticians in this field of physics, also begins with a historical overview. This concise, well-written summary gives an excellent perspective of developments dating back to the time of Rutherford.

One of the assets of Glashow's article is that we clearly see how quarks combine to give us all the baryons and mesons. The enormous, troublesome zoo of "elementary particles" that we encountered in the third and fourth papers is reduced to simple combinations of *up, down,* and *strange* quarks.

Glashow also skillfully demonstrates how the addition of a fourth quark, called *charm,* dramatically increases the total number of possible baryons and mesons. This proliferation is, however, organized and orderly. Unlike days gone by, when physicists would discover one particle after the next with apparently no rhyme or reason, Glashow illustrates the striking symmetries that emerge when charm is added to the familiar "flavors" of up, down, and strange. The beauty of this scheme is especially apparent in the diagram on page 83.

The important concept of a *gluon* is introduced by Glashow. Recall that all forces are mediated by particles that are exchanged between interacting particles. The force that glues the quarks together inside particles called hadrons involves the exchange of color. The particles that do the job have an appropriately colorful name: gluons. For example, consider just two of the quarks inside a particle such as a proton or neutron. As shown in the diagram on this page, a red up quark interacts with a blue down quark by exchanging a red-blue gluon (the wavy line).

Just as the photon is the carrier of the electromagnetic force, the gluon is thought to be the carrier of the color force. There is plenty of evidence for photons; they are simply particles or quanta of light. But gluons should also exist. They should be detectable. Unfortunately, gluons are colorful, and the theory dictates that no colorful particle can exist in isolation. Individual particles can exist only if the colors of their constituent quarks average out to white. For example, as shown on page 73 of Glashow's article, if you try to extract a colorful quark from a baryon, you expend so much energy that pairs of quarks and antiquarks are created in such a way that the emerging particle is colorless.

Since gluons are colorful, it is impossible to isolate or observe an individual gluon. Nevertheless, indirect evidence for gluons was reported during the summer of 1979. A team of physicists at the PETRA colliding-beam facility in Hamburg carefully examined the results of head-on collisions between electrons and antielectrons. According to theory, one possible result is the creation of a quark, an antiquark, and a gluon that fly away from the interaction in three different directions. These three colorful particles immediately produce quark-antiquark pairs in profusion (as in the illustration on page 73 of Glashow's article). The final result: Three jets of particles should emerge from the reaction. Three jets of particles were in fact observed.

Although there are still nonbelievers, who insist on repeating this complex experiment with greater accuracy, many prominent physicists are claiming that 1979 is the year in which gluons were discovered.

In his discussion of particle interactions, Glashow also introduces the W particle as the mediator of the weak nuclear force. Actually, as we shall see in the ninth article, by Weinberg, there are three closely related particles that do the job of the weak force. They are called W^+, W^-, and Z^0, so named because of the electric charge (positive, negative, and neutral) they carry. Whereas the strong nuclear force arises from gluons that change the color of quarks, the weak nuclear force changes the flavor of quarks via the W particles. For example, it is well known that a free neutron spontaneously decays into a proton with the emission of an electron and an antineutrino. In doing so, one of the down quarks inside the neutron is transformed into an up quark by the emission of a W^- particle. As shown in the diagram on this page, although the quark flavor has changed, its color has not. The weak force is "color-blind."

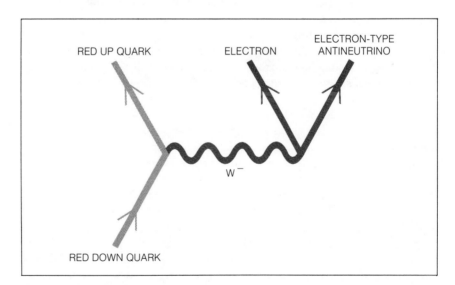

THE WEAK FORCE couples to quarks by changing flavor. In this example, a down quark (which carries charge $-\frac{1}{3}$) turns into an up quark (which carries charge $+\frac{2}{3}$) with the emission of a W^- particle (which carries charge -1). Notice that electric charge is conserved (since $-\frac{1}{3} = +\frac{2}{3} - 1$). Also notice that the quark color is unaffected. This particular reaction underlies the natural radioactive decay of free neutrons. It is the way in which a neutron (composed of two down quarks and an up quark) can turn into a proton (composed of two up quarks and a down quark).

One drawback to Glashow's article is the absence of any discussion of "anticolors." This is remedied in the next article by Nambu, which was written a year later. Quite simply, when a color is combined with its anticolor (for example, red plus antired), the result is always *white*. This is the easiest way to understand how the colors of quarks inside ordinary particles (for example, inside mesons) can average out to a colorless white. Once again, the terminology is related to color vision. However, instead of "anticolors" (commonly called "complementary colors"), the reader may be more accustomed to other names. For example, Nambu uses the additive primaries (red, blue, and green). The common names of the corresponding anticolors are:

Colors	*Anticolors*
red	antired = turquoise
blue	antiblue = yellow
green	antigreen = magenta

Nambu does a superb job of explaining how various combinations of colors and anticolors produce colorless particles, which can be detected in the laboratory. Talking about colors and anticolors is really a better way of doing things. For example, in the Feynman diagram on page 4, you should think of a red/antiblue gluon going from left to right, or (equivalently) a blue/antired gluon going from right to left.

Nambu is the first author to introduce the conept of a *gauge theory*. While he treats the topic only passing, other authors avoid it altogether. For example, in the eighth article, Johnson mentions that "it is not necessary to understand the mechanism of color transmutation in order to appreciate the relation between quark color and quark confinement." Indeed, that is correct. Accordingly, Johnson skillfully summarizes many crucial aspects of quantum chromodynamics while paying only brief lip service to the gauge approach. In fact, Johnson's 1979 article is probably the most up-to-date nonmathematical survey of quarks and color. He vividly describes *how* quantum chromodynamics works. But in order to understand *why* it works, we must appreciate the symmetries that underlie gauge theories.

Gauge theories arise from basic notions about symmetry. Symmetry involves a motion that leaves a pattern unchanged. For example, imagine an intricate tile floor in which a pattern is repeated over and over. If you move the entire tile floor by a certain amount in a certain direction, the final appearance of the floor will be indistinguishable from what you started off with. This is an example of *global symmetry*. Every point is shifted by the same amount and in the same direction.

In the ninth article, Weinberg illustrates global symmetry with a basket of apples—every apple is rotated through the same angle. In the tenth article, Freedman and Nieuwenhuizen talk about rotating a sphere (see the illustration on page 124). Instead of apples or tile floors, we should really be discussing mathematical equations that describe a physial situation. If the parameters of the equations are shifted or rotated by the same amount, and if the final equations are indistinguishable from the initial (unshifted) equations, then you have found a global symmetry.

There is a more powerful concept in physics these days, called *local symmetry*. Suppose you rotated the parameters of your equations by different amounts at different locations, like rotating each apple in Weinberg's basket through slightly different angles. If the final equations are indistinguishable from the initial equations, then you have uncovered a local symmetry. Your equations are said to be *invariant* under a local symmetry transformation.

Freedman and Nieuwenhuizen do a fine job of describing the consequences of local symmetry. As in their balloon example (see page 124), local symmetry introduces forces. Quite simply, if different parts of the balloon are rotated through slightly different angles, then the rubber becomes stretched. But at the most fundamental level, all forces are mediated by particles (see, for example, the Feynman diagrams on page 122. Consequently, the requirement of invariance under local symmetry predicts the existence of particles.

This is how gluons arise. Like the apples in Weinberg's basket, the quarks inside a particle can be imagined to have little "arrows" that point in one of three "directions," to the color red, blue, or green. By requiring invariance under local symmetry, the color force appears that is carried by the colorful gluons.

A theory that exhibits invariance under local symmetry transformations is called a gauge theory. And the particles that carry the resulting forces are called gauge particles.

This approach to physics was pioneered by C. N. Yang and Robert L. Mills in 1954. They were investigating forces between protons and neutrons inside nuclei. In particular, protons and neutrons can be considered as two aspects of the same, undifferentiated particle: the nucleon. In other words, you can imagine that a nucleon has an "arrow" that points in one of two "directions," which we call proton and neutron. Local invariance under arbitrary rotations of this "arrow" predicts the existence of three gauge particles.

In the ninth article, Weinberg elegantly describes crucial developments during the two decades following the work of Yang and Mills. In particular, Weinberg shows how these gauge particles acquire mass through "symmetry breaking." Called intermediate vector bosons (W^+, W^-, and Z^0), they mediate the weak interaction—the same process by which neutrons decay into protons.

Surely one of the most important experimental projects for the 1980s is a search for these intermediate vector bosons. Just as the photon exists (recall that the photon mediates the electromagnetic interaction), the three intermediate vector bosons should also exist. They should be real particles, just as the photon is a real particle. If W^+, W^-, and Z^0 are not found soon, then a lot of the theoretical work of the 1970s is in big trouble and everyone will have to return to the proverbial drawing board. On the other hand, the discovery of these particles would be a powerful confirmation of this approach to physics. Many physicists will be inspired to extend these methods to create a unified field theory in which *all* the forces of nature are understood as different aspects of the same basic underlying process.

The tenth and final article discusses some of these gallant attempts to formulate a unified theory. The possibility of such a theory has been a dream of physicists since the early days of Albert Einstein. For example, in 1974, Howard Georgi and Sheldon Glashow proposed a theory that unifies the strong, weak, and electromagnetic forces in much the same way that Steven Weinberg and Abdus Salam unified the weak and electromagnetic interactions. One interesting aspect of this theory (and all similar unified gauge theories based on local symmetry plus symmetry breaking) is that the proton is *not* a stable particle! For example, the proton should decay into a positron plus gamma radiation.

The half-life of the proton is very long, at least 10^{30} years. But theories like that of Georgi and Glashow predict half-lives of about 10^{32} years. Experiments are now under way to test these ideas by accurately measuring the proton's half-life. Once again, we can anticipate some answers during the early 1980s. If the proton is an unstable particle, then we are definitely on the right track to discovering some of nature's most intimate secrets.

These ideas have profound implications for the future of the universe. By observing the motions of galaxies, astronomers know that the universe is expanding. If the universe is infinite, then it will expand forever. Needless to say, the universe will be a very interesting place in 10^{32} years, when all the protons are disappearing in large numbers. Indeed, the universe is ultimately destined to be nearly devoid of matter.

This is not the first time that cosmological considerations become important in particle physics. As Johnson points out (see the table on page 102), pairs of leptons and quarks constitute families. Astrophysicists have proved that the total number of families critically affects the production of primordial helium during the few minutes following the Big Bang. Astronomical observations give an upper limit of 0.25 for the fractional abundance of this primordial helium in the universe (the remaining three-fourths is almost entirely pure hydrogen). This places an upper limit on the total number of lepton/quark families. There can be no more than eight of these families. So the process of discovering quarks (up and down, strange and charmed, top and bottom, . . . and . . .) will soon be finished.

Why has nature chosen to do this? Why does nature refuse to allow more than eight species of quarks? In fact, why does nature need more than two kinds of quarks? Most astronomers argue that they could (ideally, of course!) construct the entire universe with only the family that contains the familiar up and down quarks. So why has nature permitted the existence of other kinds of quarks? Surely not to keep physicists employed! These questions make us wonder if our understanding of the universe is perhaps still woefully incomplete.

BACKGROUND: QUANTUM MECHANICS AND PARTICLE PHYSICS

What Is Matter?

by Erwin Schrödinger
September, 1953

The wave-particle dualism afflicting modern physics is best resolved in favor of waves, believes the author, but there is no clear picture of matter on which physicists can agree

Fifty years ago science seemed on the road to a clear-cut answer to the ancient question which is the title of this article. It looked as if matter would be reduced at last to its ultimate building blocks—to certain submicroscopic but nevertheless tangible and measurable particles. But it proved to be less simple than that. Today a physicist no longer can distinguish significantly between matter and something else. We no longer contrast matter with forces or fields of force as different entities; we know now that these concepts must be merged. It is true that we speak of "empty" space (*i.e.*, space free of matter), but space is never really empty, because even in the remotest voids of the universe there is always starlight—and *that* is matter. Besides, space is filled with gravitational fields, and according to Einstein gravity and inertia cannot very well be separated.

Thus the subject of this article is in fact the total picture of space-time reality as envisaged by physics. We have to admit that our conception of material reality today is more wavering and uncertain than it has been for a long time. We know a great many interesting details, learn new ones every week. But to construct a clear, easily comprehensible picture on which all physicists would agree—that is simply impossible. Physics stands at a grave crisis of ideas. In the face of this crisis, many maintain that no objective picture of reality is possible. However, the optimists among us (of whom I consider myself one) look upon this view as a philosophical extravagance born of despair. We hope that the present fluctuations of thinking are only indications of an upheaval of old beliefs which in the end will lead to something better than the mess of formulas which today surrounds our subject.

Since the picture of matter that I am

EDITOR'S NOTE

This article is condensed from a lecture entitled "Our Conception of Matter," given by Professor Schrödinger in 1952 at a conference in Geneva organized by Rencontres Internationales de Genève. The condensation is based on a translation by Sonja Bargmann, and it is published here with the kind permission of Editions· de la Baconnière of Neuchâtel, Switzerland, who are publishing the full lecture in a volume called *L'homme devant la science*, presenting the proceedings of the conference.

supposed to draw does not yet exist, since only fragments of it are visible, some parts of this narrative may be inconsistent with others. Like Cervantes' tale of Sancho Panza, who loses his donkey in one chapter but a few chapters later, thanks to the forgetfulness of the author, is riding the dear little animal again, our story has contradictions. We must start with the well-established concept that matter is composed of corpuscles or atoms, whose existence has been quite "tangibly" demonstrated by many beautiful experiments, and with Max Planck's discovery that energy also comes in indivisible units, called quanta, which are supposed to be transferred abruptly from one carrier to another.

But then Sancho Panza's donkey will return. For I shall have to ask you to believe neither in corpuscles as permanent individuals nor in the suddenness of the transfer of an energy quantum. Discreteness is present, but not in the traditional sense of discrete single particles, let alone in the sense of abrupt processes.

Discreteness arises merely as a structure from the laws governing the phenomena. These laws are by no means fully understood; a probably correct analogue from the physics of palpable bodies is the way various partial tones of a bell derive from its shape and from the laws of elasticity to which, of themselves, nothing discontinuous adheres.

The idea that matter is made up of ultimate particles was advanced as early as the fifth century B.C. by Leucippus and Democritus, who called these particles atoms. The corpuscular theory of matter was lifted to physical reality in the theory of gases developed during the 19th century by James Clerk Maxwell and Ludwig Boltzmann. The concept of atoms and molecules in violent motion, colliding and rebounding again and again, led to full comprehension of all the properties of gases: their elastic and thermal properties, their viscosity, heat conductivity and diffusion. At the same time it led to a firm foundation of the mechanical theory of heat, namely, that heat is the motion of these ultimate particles, which becomes increasingly violent with rising temperature.

Within one tremendously fertile decade at the turn of the century came the discoveries of X-rays, of electrons, of the emission of streams of particles and other forms of energy from the atomic nucleus by radioactive decay, of the electric charges on the various particles. The masses of these particles, and of the atoms themselves, were later measured very precisely, and from this was discovered the mass defect of the atomic nucleus as a whole. The mass of a nucleus is less than the sum of the masses of its component particles; the lost mass becomes the binding energy holding the nucleus firmly together. This is called the packing effect. The nuclear forces of

LIGHT INTERFERENCE pattern, showing the wave nature of light, was produced at the National Bureau of Standards, using light from mercury vapor and an interferometer.

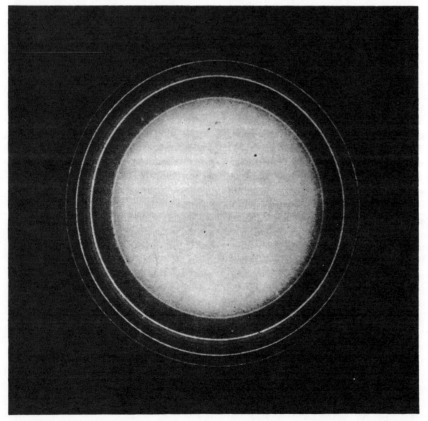

ELECTRON INTERFERENCE pattern from a crystal diffraction experiment at the Radio Corporation of America Laboratories gives convincing evidence that electrons are waves.

course are not electrical forces—those are repellent—but are much stronger and act only within very short distances, about 10^{-13} centimeter.

Here I am already caught in a contradiction. Didn't I say at the beginning that we no longer assume the existence of force fields apart from matter? I could easily talk myself out of it by saying: Well, the force field of a particle is simply considered a part of it. But that is not the fact. The established view today is rather that everything is at the same time both particle and field. Everything has the continuous structure with which we are familiar in fields, as well as the discrete structure with which we are equally familiar in particles. This concept is supported by innumerable experimental facts and is accepted in general, though opinions differ on details, as we shall see.

In the particular case of the field of nuclear forces, the particle structure is more or less known. Most likely the continuous force field is represented by the so-called pi mesons. On the other hand, the protons and neutrons, which we think of as discrete particles, indisputably also have a continuous wave structure, as is shown by the interference patterns they form when diffracted by a crystal. The difficulty of combining these two so very different character traits in one mental picture is the main stumbling-block that causes our conception of matter to be so uncertain.

Neither the particle concept nor the wave concept is hypothetical. The tracks in a photographic emulsion or in a Wilson cloud chamber leave no doubt of the behavior of particles as discrete units. The artificial production of nuclear particles is being attempted right now with terrific expenditure, defrayed in the main by the various state ministries of defense. It is true that one cannot kill anybody with one such racing particle, or else we should all be dead by now. But their study promises, indirectly, a hastened realization of the plan for the annihilation of mankind which is so close to all our hearts.

You can easily observe particles yourself by looking at a luminous numeral of your wrist watch in the dark with a magnifying glass. The luminosity surges and undulates, just as a lake sometimes twinkles in the sun. The light consists of sparklets, each produced by a so-called alpha particle (helium nucleus) expelled by a radioactive atom which in this process is transformed into a different atom. A specific device for detecting and recording single particles is the

Geiger-Müller counter. In this short ré-
sumé I cannot possibly exhaust the many
ways in which we can observe single
particles.

Now to the continuous field or wave
character of matter. Wave structure
is studied mainly by means of diffraction
and interference—phenomena which occur when wave trains cross each other.
For the analysis and measurement of
light waves the principal device is the
ruled grating, which consists of a great
many fine, parallel, equidistant lines,
closely engraved on a specular metallic
surface. Light impinging from one direction is scattered by them and collected in different directions depending
on its wavelength. But even the finest
ruled gratings we can produce are too
coarse to scatter the very much shorter
waves associated with matter. The fine
lattices of crystals, however, which Max
von Laue first used as gratings to analyze
the very short X-rays, will do the same
for "matter waves." Directed at the surface of a crystal, high-velocity streams
of particles manifest their wave nature.
With crystal gratings physicists have diffracted and measured the wavelengths
of electrons, neutrons and protons.

What does Planck's quantum theory
have to do with all this? Planck told us
in 1900 that he could comprehend the
radiation from red-hot iron, or from an
incandescent star such as the sun, only
if this radiation was produced in discrete
portions and transferred in such discrete
quantities from one carrier to another
(*e.g.*, from atom to atom). This was extremely startling, because up to that time
energy had been a highly abstract concept. Five years later Einstein told us
that energy has mass and mass is energy; in other words, that they are one
and the same. Now the scales begin to
fall from our eyes: our dear old atoms,
corpuscles, particles are Planck's energy
quanta. *The carriers of those quanta are
themselves quanta.* One gets dizzy.
Something quite fundamental must lie
at the bottom of this, but it is not surprising that the secret is not yet understood. After all, the scales did not fall
suddenly. It took 20 or 30 years. And
perhaps they still have not fallen completely.

The next step was not quite so far-
reaching, but important enough. By an
ingenious and appropriate generaliza-
tion of Planck's hypothesis Niels Bohr
taught us to understand the line spectra
of atoms and molecules and how atoms
were composed of heavy, positively

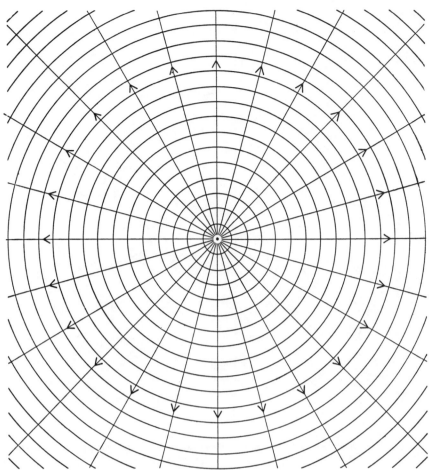

WAVE DIAGRAM in two dimensions shows wave fronts (*circles*) and wave "normals" or
"rays" (*arrows*). In three dimensions the fronts would be surfaces like layers in an onion.

charged nuclei with light, negatively
charged electrons revolving around
them. Each small system—atom or molecule—can harbor only definite discrete
energy quantities, corresponding to its
nature or its constitution. In transition
from a higher to a lower "energy level"
it emits the excess energy as a radiation
quantum of definite wavelength, inversely proportional to the quantum
given off. This means that a quantum of
given magnitude manifests itself in a
periodic process of definite frequency
which is directly proportional to the
quantum; the frequency equals the energy quantum divided by the famous
Planck's constant, h.

According to Einstein a particle has
the energy mc^2, m being the mass of the
particle and c the velocity of light. In
1925 Louis de Broglie drew the infer-
ence, which rather suggests itself, that
a particle might have associated with
it a wave process of frequency mc^2 divided by h. The particle for which he
postulated such a wave was the electron. Within two years the "electron
waves" required by his theory were dem-

onstrated by the famous electron diffraction experiment of C. J. Davisson
and L. H. Germer. This was the starting
point for the cognition that everything—
anything at all—is simultaneously particle and wave field. Thus de Broglie's
dissertation initiated our uncertainty
about the nature of matter. Both the particle picture and the wave picture have
truth value, and we cannot give up
either one or the other. But we do not
know how to combine them.

That the two pictures are connected
is known in full generality with great
precision and down to amazing details.
But concerning the unification to a single, concrete, palpable picture opinions
are so strongly divided that a great many
deem it altogether impossible. I shall
briefly sketch the connection. But do not
expect that a uniform, concrete picture
will emerge before you; and do not
blame the lack of success either on my
ineptness in exposition or your own
denseness—nobody has yet succeeded.

One distinguishes two things in a
wave. First of all, a wave has a front,

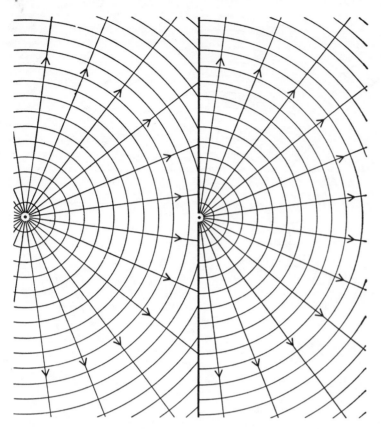

DIFFRACTION is characteristic of waves. When a wave (*left*) comes to a barrier perforated with a small hole, it diffracts around the edges of the hole to form a new wave (*right*).

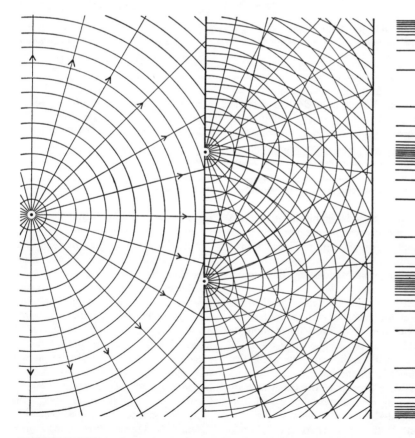

INTERFERENCE is also evidence of waves. Its characteristic pattern is formed when rays interact. For light waves the pattern shows up as bright and dark bands on a screen (*right*).

and a succession of wave fronts forms a system of surfaces like the layers of an onion. You are familiar with the two-dimensional analogue of the beautiful wave circles that form on the smooth surface of a pond when a stone is thrown in. The second characteristic of a wave, less intuitive, is the path along which it travels—a system of imagined lines perpendicular to the wave fronts. These lines are known as the wave "normals" or "rays."

We can make the provisional assertion that these rays correspond to the trajectories of particles. Indeed, if you cut a small piece out of a wave, approximately 10 or 20 wavelengths along the direction of propagation and about as much across, such a "wave packet" would actually move along a ray with exactly the same velocity and change of velocity as we might expect from a particle of this particular kind at this particular place, taking into account any force fields acting on the particle.

Here I falter. For what I must say now, though correct, almost contradicts this provisional assertion. Although the behavior of the wave packet gives us a more or less intuitive picture of a particle, which can be worked out in detail (*e.g.*, the momentum of a particle increases as the wavelength decreases; the two are inversely proportional), yet for many reasons we cannot take this intuitive picture quite seriously. For one thing, it is, after all, somewhat vague, the more so the greater the wavelength. For another, quite often we are dealing not with a small packet but with an extended wave. For still another, we must also deal with the important special case of very small "packelets" which form a kind of "standing wave" which can have no wave fronts or wave normals.

One interpretation of wave phenomena which is extensively supported by experiments is this: At each position of a uniformly propagating wave train there is a twofold structural connection of interactions, which may be distinguished as "longitudinal" and "transversal." The transversal structure is that of the wave fronts and manifests itself in diffraction and interference experiments; the longitudinal structure is that of the wave normals and manifests itself in the observation of single particles. However, these concepts of longitudinal and transversal structures are not sharply defined and absolute, since the concepts of wave front and wave normal are not, either.

The interpretation breaks down completely in the special case of the standing waves mentioned above. Here the whole wave phenomenon is reduced to a small region of the dimensions of a single or very few wavelengths. You can produce standing water waves of a similar nature in a small basin if you dabble with your finger rather uniformly in its center, or else just give it a little push so that the water surface undulates. In this situation we are not dealing with uniform wave propagation; what catches the interest are the normal frequencies of these standing waves. The water waves in the basin are an analogue of a wave phenomenon associated with electrons, which occurs in a region just about the size of the atom. The normal frequencies of the wave group washing around the atomic nucleus are universally found to be exactly equal to Bohr's atomic "energy levels" divided by Planck's constant h. Thus the ingenious yet somewhat artificial assumptions of Bohr's model of the atom, as well as of the older quantum theory in general, are superseded by the far more natural idea of de Broglie's wave phenomenon. The wave phenomenon forms the "body" proper of the atom. It takes the place of the individual pointlike electrons which in Bohr's model are supposed to swarm around the nucleus. Such pointlike single particles are completely out of the question within the atom, and if one still thinks of the nucleus itself in this way one does so quite consciously for reasons of expediency.

What seems to me particularly important about the discovery that "energy levels" are virtually nothing but the frequencies of normal modes of vibration is that now one can do without the assumption of sudden transitions, or quantum jumps, since two or more normal modes may very well be excited simultaneously. The discreteness of the normal frequencies fully suffices—so I believe—to support the considerations from which Planck started and many similar and just as important ones—I mean, in short, to support all of quantum thermodynamics.

The theory of quantum jumps is becoming more and more inacceptable, at least to me personally, as the years go on. Its abandonment has, however, far-reaching consequences. It means that one must give up entirely the idea of the exchange of energy in well-defined quanta and replace it with the concept of resonance between vibrational frequencies. Yet we have seen that because of the identity of mass and energy, we

must consider the particles themselves as Planck's energy quanta. This is at first frightening. For the substituted theory implies that we can no longer consider the individual particle as a well-defined permanent entity.

That it is, in fact, no such thing can be reasoned in other ways. For one thing, there is Werner Heisenberg's famous uncertainty principle, according to which a particle cannot simultaneously have a well-defined position and a sharply defined velocity. This uncertainty implies that we cannot be sure that the same particle could ever be observed twice. Another conclusive reason for not attributing identifiable sameness to individual particles is that we must obliterate their individualities whenever we consider two or more interacting particles of the same kind, *e.g.*, the two electrons of a helium atom. Two situations which are distinguished only by the interchange of the two electrons must be counted as one and the same; if they are counted as *two* equal situations, nonsense obtains. This circumstance holds for any kind of particle in arbitrary numbers without exception.

Most theoreticians will probably accept the foregoing reasoning and admit that the individual particle is not a well-defined permanent entity of detectable identity or sameness. Nevertheless this inadmissible concept of the individual particle continues to play a large role in their ideas and discussions. Even deeper rooted is the belief in "quantum jumps," which is now surrounded with a highly abstruse terminology whose common-sense meaning is often difficult to grasp. For instance, an important word in the standing vocabulary of quantum theory is "probability," referring to transition from one level to another. But, after all, one can speak of the probability of an event only assuming that, occasionally, it actually occurs. If it does occur, the transition must indeed be sudden, since intermediate stages are disclaimed. Moreover, if it takes time, it might conceivably be interrupted halfway by an unforeseen disturbance. This possibility leaves one completely at sea.

The wave *v.* corpuscle dilemma is supposed to be resolved by asserting that the wave field merely serves for the computation of the probability of finding a particle of given properties at a given position if one looks for it there. But once one deprives the waves of reality and assigns them only a kind of informative role, it becomes very difficult to under-

stand the phenomena of interference and diffraction on the basis of the combined action of discrete single particles. It certainly seems easier to explain particle tracks in terms of waves than to explain the wave phenomenon in terms of corpuscles.

"Real existence" is, to be sure, an expression which has been virtually chased to death by many philosophical hounds. Its simple, naive meaning has almost become lost to us. Therefore I want to recall something else. I spoke of a corpuscle's not being an individual. Properly speaking, one never observes the same particle a second time—very much as Heraclitus says of the river. You cannot mark an electron, you cannot paint it red. Indeed, you must not even *think* of it as marked; if you do, your "counting" will be false and you will get wrong results at every step—for the structure of line spectra, in thermodynamics and elsewhere. A wave, on the other hand, can easily be imprinted with an individual structure by which it can be recognized beyond doubt. Think of the beacon fires that guide ships at sea. The light shines according to a definite code; for example: three seconds light, five seconds dark, one second light, another pause of five seconds, and again light for three seconds—the skipper knows that is San Sebastian. Or you talk by wireless telephone with a friend across the Atlantic; as soon as he says, "Hello there, Edward Meier speaking," you know that his voice has imprinted on the radio wave a structure which can be distinguished from any other. But one does not have to go that far. If your wife calls, "Francis!" from the garden, it is exactly the same thing, except that the structure is printed on sound waves and the trip is shorter (though it takes somewhat longer than the journey of radio waves across the Atlantic). All our verbal communication is based on imprinted individual wave structures. And, according to the same principle, what a wealth of details is transmitted to us in rapid succession by the movie or the television picture!

This characteristic, the individuality of the wave phenomenon, has already been found to a remarkable extent in the very much finer waves of particles. One example must suffice. A limited volume of gas, say helium, can be thought of either as a collection of many helium atoms or as a superposition of elementary wave trains of matter waves. Both views lead to the same theoretical results as to the behavior of the gas upon heating, compression, and so on. But

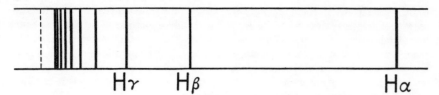

HYDROGEN SPECTRUM expresses the behavior of a fundamental constituent of matter, the electron. Shown above is a part of the Balmer series of spectral lines, which are in the visible light range. Each line is the result of a change in energy of the atom's electron.

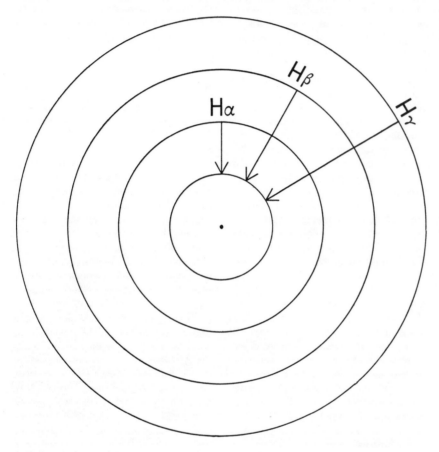

BOHR THEORY explained spectral lines of hydrogen by postulating a pointlike electron revolving around the nucleus in any of a number of possible orbits. In falling from one to another, the electron emits light energy whose wavelength is that of one of the spectral lines.

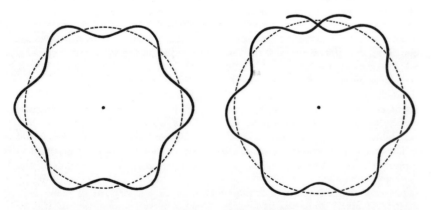

WAVE MECHANICS sees the electron not as a point mass, but as a standing wave washing to and fro in the atom. Some modes of vibration are possible (left), while others are not (right). The possible modes correspond exactly to the Bohr theory's possible energy levels.

when you attempt to apply certain somewhat involved enumerations to the gas, you must carry them out in different ways according to the mental picture with which you approach it. If you treat the gas as consisting of particles, then no individuality must be ascribed to them, as I said. If, however, you concentrate on the matter wave trains instead of on the particles, every one of the wave trains has a well-defined structure which is different from that of any other. It is true that there are many pairs of waves which are so similar to each other that they could change roles without any noticeable effect on the gas. But if you should count the very many similar states formed in this way as merely a single one, the result would be quite wrong.

In spite of everything we cannot completely banish the concepts of quantum jump and individual corpuscle from the vocabulary of physics. We still require them to describe many details of the structure of matter. How can one ever determine the weight of a carbon nucleus and of a hydrogen nucleus, each to the precision of several decimals, and detect that the former is somewhat lighter than the 12 hydrogen nuclei combined in it, without accepting for the time being the view that these particles are something quite concrete and real? This view is so much more convenient than the roundabout consideration of wave trains that we cannot do without it, just as the chemist does not discard his valence-bond formulas, although he fully realizes that they represent a drastic simplification of a rather involved wave-mechanical situation.

If you finally ask me: "Well, what *are* these corpuscles, really?" I ought to confess honestly that I am almost as little prepared to answer that as to tell where Sancho Panza's second donkey came from. At the most, it may be permissible to say that one can think of particles as more or less temporary entities within the wave field whose form and general behavior are nevertheless so clearly and sharply determined by the laws of waves that many processes take place *as if* these temporary entities were substantial permanent beings. The mass and the charge of particles, defined with such precision, must then be counted among the structural elements determined by the wave laws. The conservation of charge and mass in the large must be considered as a statistical effect, based on the "law of large numbers."

Field Theory

by Freeman J. Dyson

April, 1953

*The physicist speaks of classical fields and quantum
fields. Exactly what are they, and what is their role in
modern physics and in our present view of reality?*

"IT IS perhaps surprising that no new
meson was reported during the sym-
posium, though almost a month had
passed since a previous meeting of nuclear physicists in Copenhagen."

This learned joke in the British journal
Nature, commenting on an international
physics conference last summer, sums
up very well the present chaotic situation in theoretical physics. We have become accustomed during the last few
years to the discovery of new particles.
About 20 different kinds are now known.
Everybody expects that many more will
be discovered as experimental techniques are improved. Yet nobody has
had any success in classifying the known
particles, or in predicting the properties
of unknown ones. Nobody understands
why such and such particles exist, why
they have the particular masses that are
observed or why some of them strongly
interact and some do not.

How do the theoretical physicists
spend their time, if they are not able to
attack the fundamental problem of the
nature of elementary particles? Of what
use can the existing atomic theories be,
if they do not throw light on this basic
problem? These awkward questions are
asked rather frequently when experimental and theoretical physicists come
together. I shall try to answer them and
to explain why we theoretical physicists
believe that our theories are useful even
though there is so much we do not
understand.

First it is necessary to make one point
clear: there *is* an official and generally
accepted theory of elementary particles,
known as the "quantum field theory."
While theoretical physicists often dis-
agree about the finer details of the
theory, and especially about the way in
which it should be applied to practical
problems, the great majority of them
agree that the theory in its main features
is correct. The minority who reject the
theory, although led by the great names
of Albert Einstein and P. A. M. Dirac,
do not yet have any workable alternative
to put in its place. In this article I shall

adopt the point of view of the majority.
When I talk about the concept of field,
I mean specifically the concept as it is
used in the present-day official quantum
field theory. The majority believes that

this concept is so useful and illuminat-
ing that it will survive the changes and
revolutions which the theory will inevitably undergo in the future. Hence-
forward I shall omit the phrase "in the

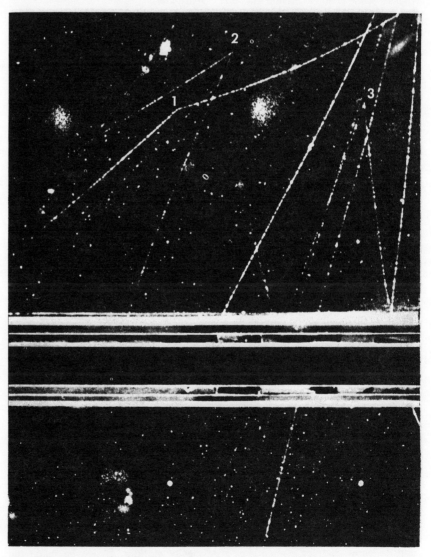

NEUTRAL V-PARTICLES gave rise to the three V-shaped tracks in this
cloud-chamber photograph by R. B. Leighton of the California Institute of
Technology. Below the center of the photograph is the edge of a lead plate.

opinion of the majority," or "in my opinion," which strictly ought to stand at the beginning of every sentence.

A Descriptive Theory

It is important to make a second general remark about the theory at the outset. This concerns the failure of the theory to give us an understanding of why the observed elementary particles exist and no others. The point is that the theory is in its nature descriptive and not explanatory. It describes how elementary particles behave; it does not attempt to explain why they behave so. To draw an analogy from a familiar branch of science, the function of chemistry as it existed before 1900 was to describe precisely the properties of the chemical elements and their interactions. Chemistry described how the elements behave; it did not try to explain why a particular set of elements, each with its particular properties, exists. To answer the question "why," completely new sciences were needed: atomic and nuclear physics. Looking backward, it is now clear that 19th-century chemists were right to concentrate on the "how" and to ignore the "why." They did not have the tools to begin to discuss intelligently the reasons for the individualities of the elements. They had to spend a hundred years building up a good quantitative descriptive theory before they could go further. And the result of their labors—the classical science of chemistry—was not destroyed or superseded by the later insight that atomic physics gave.

The quantum field theory treats elementary particles just as 19th-century chemists treated the elements. The theory starts from the existence of a specified list of elementary particles, with specified masses, spins, charges and specified interactions with one another. All these data are put into the theory at the beginning. The purpose of the theory is simply to deduce from this information what will happen if particle A is fired at particle B with a given velocity. We are not yet sure whether the theory will be able to fulfill even this modest purpose completely. Many technical difficulties have still to be overcome. One of the difficulties is that we do not yet have the complete list of elementary particles. Nevertheless the successes of the theory in describing experimental results have been striking. It seems likely that the theory in something like its present form will describe accurately a very wide range of possible experiments. This is the most that we would wish to claim for it.

Our justification for concentrating attention so heavily on the existing theory, with its many arbitrary assumptions, is the belief that a working descriptive theory of elementary particles must be established before we can expect to reach a more complete understanding at a deeper level. The numerous attempts to by-pass the historical process, and to understand the elementary particles on the basis of general principles without waiting for a descriptive theory, have been as unsuccessful as they were ambitious. In fact, the more ambitious they are, the more unsuccessful. These attempts seem to be on a level with the famous 19th-century attempts to explain atoms as "vortices in the ether."

Classical Fields

Physicists talk about two kinds of fields: classical fields and quantum fields. Actually we believe that all fields in nature are quantum fields. A classical field is just a special large-scale manifestation of a quantum field. But since classical fields were discovered first and are easier to understand, it is necessary to say what we mean by a classical field first, and go on to talk about quantum fields later.

A classical field is a kind of tension or stress which can exist in empty space in the absence of matter. It reveals itself by producing forces, which act on any material objects that happen to lie in the space the field occupies. The standard examples of classical fields are the electric and magnetic fields, which push and pull electrically charged objects and magnetized objects respectively. Michael Faraday discovered that these two fields also exert effects on each other. He found that a changing magnetic field produces electric forces (an effect now known as induction), and his finding made possible the development of practical electric generators. Later the exact laws of behavior of electric and magnetic fields were formulated mathematically by James Clerk Maxwell. He found that in any space where a changing magnetic field exists, an electric field must exist also, and vice versa. In order to describe completely the state of the fields in a given region of space, it is necessary to specify the strength and the direction of both the electric and magnetic fields at every point of the region separately. This is the characteristic mathematical property of a classical field: it is an undefined something which exists throughout a volume of space and which is described by sets of numbers, each set denoting the field strength and direction at a single point in the space.

Maxwell was the first to realize that electric and magnetic fields could exist not only near charges and magnets but also in free space completely disconnected from material objects. From his equations he deduced that in empty space such fields would travel with the velocity of light. Hence he made the epoch-making guess that light consists of traveling electromagnetic fields. We

NAME	SYMBOL
PHOTON	γ
GRAVITON	G
NEUTRINO	ν
ELECTRON	e
POSITRON	p
POSITIVE MU MESON	μ^+
NEGATIVE MU MESON	μ^-
NEUTRAL PI MESON	π^0
POSITIVE PI MESON	π^+
NEGATIVE PI MESON	π^-
ZETA MESON?	ζ
NEUTRAL V-PARTICLE	V_2^0
TAU MESON	τ
KAPPA MESON	κ
POSITIVE CHI MESON	χ^+
NEGATIVE CHI MESON	χ^-
PROTON	P
NEUTRON	N
NEUTRAL V-PARTICLE	V_1^0
POSITIVE V-PARTICLE?	V^+

CHART of the fundamental particles has been revised since a similar chart appeared in this magazine for

now know that his guess was correct, and we are even able to manufacture traveling electromagnetic fields ourselves and use them for various purposes. These artificial traveling fields we call radio.

Another example of a classical field is the gravitational field. This has the special property that it acts on all material objects in a given region of space. It is very difficult to experiment with, because the gravitational field produced by any object of convenient laboratory size is absurdly weak. For this reason we have never been able to detect any effects of freely traveling gravitational waves, which presumably exist in the neighborhood of a rapidly oscillating mass. It is also impossible to measure any possible interactions of the gravita-

HARGE	MASS	SPIN	STATISTICS	LIFETIME (SECONDS)	DECAY SCHEME
0	0	1	BOSE-EINSTEIN	STABLE	
0	0	2	BOSE-EINSTEIN	STABLE	
0	0	½	FERMI-DIRAC	STABLE	
—	1	½	FERMI-DIRAC	STABLE	
+	1	½	FERMI-DIRAC	STABLE	
+	210	½	FERMI-DIRAC	2.1×10^{-6}	$\mu^+ \to p + 2\nu$
—	210	½	FERMI-DIRAC	2.1×10^{-6}	$\mu^- \to e + 2\nu$
0	265	0	BOSE-EINSTEIN	10^{-15}	$\pi^0 \to 2\gamma$
+	276	0	BOSE-EINSTEIN	2.6×10^{-8}	$\pi^+ \to \mu^+ + \nu$
—	276	0	BOSE-EINSTEIN	2.6×10^{-8}	$\pi^- \to \mu^- + \nu$
±	550	?	?	10^{-12}	$\zeta \to \pi + ?$
0	850	?	?	10^{-10}	$V_2^0 \to \pi^+ + \pi^- + ?$
±	975	?	BOSE-EINSTEIN	10^{-8}	$\tau \to 3\pi$
±	1100	?	?	?	$\kappa \to \mu + ?$
+	1400	?	?	10^{-9}	$\chi^+ \to \pi^+ + ?$
—	1400	?	?	10^{-9}	$\chi^- \to \pi^- + ?$
+	1836	½	FERMI-DIRAC	STABLE	
0	1838.5	½	FERMI-DIRAC	750	$N \to P + e + \nu$
0	2190	?	FERMI-DIRAC	3×10^{-10}	$V_1^0 \to P + \pi^-$
+	2200	?	?	10^{-9}	$V^+ \to P + ?$

January, 1952. Among the changes are the addition of the chi mesons and a second variety of neutral V-particle. The particles are listed in the order of their mass. The existence of the zeta meson and the positive V-particle is doubtful. The masses and the lifetimes of all the newer particles listed in the chart are approximate.

tional and electromagnetic fields. This is most unfortunate, and it is the main reason why we know so much less about gravitation than about the other fields.

A Model of a Field

What, then, is the picture we have in mind when we try to visualize a classical field? Characteristically, modern physicists do not seriously try to visualize the objects they discuss. In the 19th century it was different. Then it seemed that the universe was built of solid mechanical objects, and that to understand an electric field it was necessary to visualize the field as a mechanical stress in a material substance. It was possible, indeed, to visualize electric and magnetic fields in this way. To do so men imagined a mate-

rial substance called the ether, which was supposed to fill the whole of space and carry the electric and magnetic stresses. But as the theory was developed, the properties of the ether became more and more extraordinary and self-contradictory. Einstein in 1905 finally abandoned the ether and proposed a new and simple version of the Maxwell theory in which the ether was never once mentioned. Since 1905 the idea that everything in the universe should be visualized mechanically has gradually become ridiculous. We now find that mechanical objects themselves are composed of atoms held together by electric fields, and therefore it makes no sense to try to explain the electric fields in terms of mechanical models.

It is still convenient sometimes to

make a mental picture of an electric field. For example, we may think of it as a flowing liquid which fills a given space and which at each point has a certain velocity (strength) and direction of flow. But nobody nowadays imagines that the liquid really exists or that it explains the behavior of the field. The flowing liquid is just a model—a convenient way to express our knowledge about the field in concrete terms. It is a good model only so long as we remember not to take it too seriously. We must not, for example, expect that the equations of motions of the electric field will be the same as those of any self-respecting liquid. To a modern physicist the electric field is a fundamental concept which cannot be reduced to anything simpler. It is a unique something with a

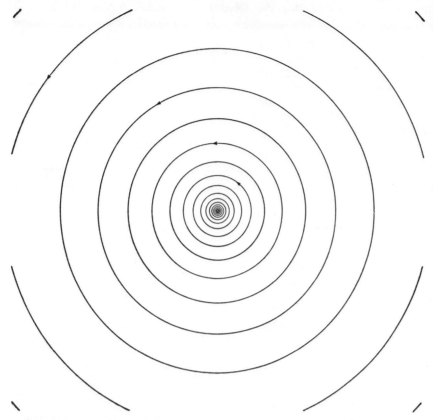

CLASSICAL FIELD is schematically depicted as a series of concentric rings around a point. These rings might show the magnetic field around an electric current traveling along a wire perpendicular to the surface of this page.

set of known properties, and that is all there is to it. This being understood, the reader may safely think of the flowing liquid as a fairly accurate representation of what we mean by a classical electric field. The electric and magnetic fields must then be pictured as two different liquids, both filling the whole of space, moving separately and interpenetrating each other freely. At each point there are two velocities, representing the strengths of the electric and magnetic components of the total electromagnetic field.

It is characteristic of a classical field that its strength at a given point varies smoothly as the point moves around in space. Therefore the liquid model must be imagined as an ideal liquid, not composed of atoms but filling all space uniformly and having a well-defined velocity at every point.

The new idea that Einstein introduced in 1905, and that killed the ether, was the principle of relativity. This principle states that the properties of empty space are always the same, regardless of the velocity with which an experimenter is moving through it. Thus even if there is a material ether filling space, the experimenter is unable to measure the velocity of himself relative to it; for all practical purposes the ether is unobservable. All that we can certainly say about it is that if it does exist, it is of no interest to us. Our picture of the world

becomes much simpler if we abandon the ether and speak only about electric and magnetic fields in empty space.

Einstein made a complete theory of the classical electromagnetic field and its interactions with matter, using the principle of relativity as his starting point. In 1916 he extended the idea of relativity to construct his theory of the classical gravitational field. These theories stand today substantially as Einstein left them.

The classical field theories of Einstein —electromagnetic and gravitational—together give us a satisfactory explanation of all large-scale physical phenomena. That is to say, they explain everything in the physical world that can be explained without bringing into view the fact that the world is built of elementary particles. There is every reason to believe that the classical field theories are correct so long as we are talking about objects much bigger and heavier than a single atom. But they fail completely to describe the behavior of individual atoms and particles. To understand the small-scale side of physics, physicists had to invent quantum mechanics and the idea of a quantum field.

Quantum Fields

Unfortunately the quantum field is even more difficult to visualize than the classical field. The basic axiom of quantum mechanics is the uncertainty prin-

ciple. This says that the more closely we look at any object, the more the object is disturbed by our looking at it, and the less we can know about the subsequent state of the object. Another less precise way of expressing the same principle is this: All objects of atomic size fluctuate continually; they cannot maintain a precisely defined position for a finite length of time. Their quantum fluctuations are never precisely predictable, and the laws of quantum mechanics tell us only the statistical behavior of the fluctuations when averaged over a long time. The universal existence of these fluctuations, and the general correctness of the laws of quantum mechanics, have been verified by a wealth of experiments during the last 30 years.

How do the quantum fluctuations affect the classical field? The answer is: not at all. The fluctuations are not observable with any ordinary large-scale equipment, for they average out to produce no effect on these instruments. Looked at with large-scale apparatus, the quantum field behaves exactly like a classical field. Only when we measure the effects of an electromagnetic field on a single atom do the quantum fluctuations of the field become noticeable.

The physicists Willis Lamb and Robert C. Retherford at Columbia University have observed the effects of electromagnetic fields on single hydrogen atoms with a piece of apparatus known to radar experts as a microwave cavity resonator [see "Radio Waves and Matter," by Harry M. Davis; SCIENTIFIC AMERICAN, September, 1948]. Using the techniques of microwave spectroscopy, they were able to measure the effects of the fields with great accuracy. The effect of the quantum fluctuations, itself a small part of the total effect of the fields, was measured to an accuracy better than one part in a thousand, and within this margin of possible error the effect agreed with the conclusions of the quantum field theory. The Lamb-Retherford experiment is the strongest evidence we have for believing that our picture of the quantum field is correct in detail.

At the risk of making some professional quantum theoreticians turn pale, I shall describe a mechanical model which may give some idea of the nature of a quantum field. Imagine the flowing liquid which served as a model for a classical electric field. But suppose that the flow, instead of being smooth, is turbulent, like the wake of an ocean liner. Superimposed on the steady average motion there is a tremendous confusion of eddies, of all different sizes and overlapping and mingling with one another. In any small region of the liquid the velocity continually fluctuates, in a more or less random way. The smaller the region, the wilder and more rapid are the velocity fluctuations. In a real liquid these fluctuations are finally limited by two factors: (1) the viscosity, or stickiness, of the liquid, which damps out the turbulent motions, and (2) the atomic structure of the liquid, which sets a minimum size for the eddies, since it is meaningless to talk about eddies containing only a few atoms. In our model of the quantum field, however, we assume that neither of these factors operates. There is no dissipation of energy by viscosity, or any minimum size of eddies. Consequently the velocity in a given region can continue to fluctuate without diminution forever, and the fluctuations grow more and more intense without limit as the size of the region is reduced.

The model does not describe correctly the detailed quantum-mechanical properties of a quantum field; no classical model can do that. But it does seem to me to give a reasonably valid picture of the general appearance of the thing. In particular, the model makes clear that it is strictly meaningless to speak about the velocity of the liquid at any specific point. The fluctuations in the neighborhood of the point become infinitely large as the neighborhood becomes smaller, and so the velocity at the point itself has no meaning. The only quantities that have meaning are averaged velocities, taken over a given region of space and over a given time interval. This property of the model is a true representation of a property of a quantum field. The strength of a quantum field at a point can never be measured. The whole quantum field theory is a theory of the behavior of field strengths averaged over finite regions of space and time.

The Particles Emerge

Now comes the climax of the story. We have put into the theory of the quantum field two big ideas: the idea of quantum mechanics and the idea of relativity. These two ideas force us to construct a mathematical theory which in its main lines is fixed; the only freedom left to us is in matters of detail. When we deduce the consequences of this mathematical theory, we find that a miracle has occurred: automatically there emerges a third big idea—that the world is built of elementary particles.

This idea is a consequence of the fact that in a quantum field energy can exist only in discrete units, which we call quanta. When we work out the theory of these quanta in detail, we find that they have precisely the properties of the elementary particles that we observe in the world around us.

It is not possible, in an article such as this, to explain how the elementary particles arise mathematically out of the fluctuations of a field. It cannot be understood by thinking about turbulent liquids or any classical model. All I can say here is that it happens. And it is the basic permanent reason for believing that the concept of a quantum field is a valid concept and will survive any changes that may later be made in matters of detail.

The picture of the world that we have finally reached is the following: Some 10 or 20 qualitatively different quantum fields exist. Each fills the whole of space and has its own particular properties. There is nothing else except these fields; the whole of the material universe is built of them. Between various pairs of the fields there are various kinds of interaction. Each field manifests itself as a type of elementary particle. The particles of a given type are always completely identical and indistinguishable. The number of particles of a given type is not fixed, for particles are constantly being created or annihilated or transmuted into one another. The properties of the interactions determine the rules for creation and transmutation of particles.

In this picture of the world the electromagnetic field appears on an exactly equal footing with the other fields. The particle corresponding to it is the light quantum, or photon. The photon appears to be different from other elementary particles only because its laws of interaction make it especially easy to create and annihilate. So the photon appears to be less permanent than, for example, the electron. But this is only a difference of degree; all particles, including the electron, can be rapidly annihilated under suitable conditions.

The elementary particle corresponding to the gravitational field has been named the graviton. There can be little doubt that in a formal mathematical sense the graviton exists. However, nobody has ever observed an individual graviton. Because of the extreme weakness of the gravitational interaction, in practice only large masses produce observable gravitational effects. In the case of large masses, the number of gravitons involved in the interaction is very large, and the field behaves like a classical field. Consequently, many physicists believe that the individual graviton never will be observed. Whether the graviton has a real existence is one of the most important open questions in physics.

The electromagnetic and gravitational fields have one essential property in common. They are long-range fields which make their effects felt over great distances. This is connected with the fact that the photon and the graviton are particles which have no rest-mass and always travel at a fixed velocity—the velocity of light. Almost all other fields in nature have a short range, less than the size of an atom, and their effects cannot be felt beyond this distance. The short-range fields cannot be detected in a classical way by measuring their effects on large objects. They never behave like classical fields in any experimental situation. This is why, for example, the field corresponding to the electron was never recognized as a field until the quantum field theory was developed. And even now the electron field seems more peculiar and foreign to us than the electromagnetic field. Fundamentally the two are very similar. The main difference between them is the short range of the electron field, which has the consequence that the electron possesses a rest-mass and can travel with any velocity not exceeding the velocity of light. Most of the other known particles—protons, neutrons, the many varieties of mesons—also have a rest-mass and are associated with short-range fields.

Positive and Negative

Perhaps the most spectacular success of the quantum field theory is in its treatment of charged fields. According to the theory, a quantum field may or may not carry an electric charge. For example, the electron field carries a charge, while the electromagnetic field does not. The theory automatically predicts that any charged field must be represented by two types of particle, precisely alike in all respects except that one has a positive charge and the other negative. The theory also predicts that under suitable conditions a pair of such particles, one positively and one negatively charged, can be created or annihilated together in a single event. All these predictions of the theory have been completely confirmed in the case of the electron field. There exists a particle, the positron, which is exactly like an electron except that it has the opposite charge. It has also been proved that there are at least two varieties of meson that exist in positive and negative forms. The theory predicts that there should be an antiproton: a particle negatively charged but otherwise identical with a proton. The antiproton has not yet been detected. It presents an outstanding challenge to experimental physicists to discover it, or to theoretical physicists to explain why it should not exist.

Even to a hardened theoretical physicist it remains perpetually astonishing that our solid world of trees and stones can be built of quantum fields and nothing else. The quantum field seems far too fluid and insubstantial to be the basic stuff of the universe. Yet we have learned gradually to accept the fact that the laws of quantum mechanics impose their own peculiar rigidity upon the fields they govern, a rigidity which is alien to our intuitive conceptions but which nonetheless effectively holds the earth in place. We have learned to apply, both to ourselves and to our subject, the words of Robert Bridges:

*Our stability is but balance, and
our wisdom lies
In masterful administration of the
unforeseen.*

3

Elementary Particles

by Murray Gell-Mann and E. P. Rosenbaum
July, 1957

An account of the abstract theoretical ideas which physicists use to help them understand the material world. These ideas begin to show some order in the jumble of subatomic particles

There is no excellent beauty that hath not some strangeness in the proportion.
—Francis Bacon

This aphorism quoted from Bacon is doubtless true of science as well as art. But is too much strangeness not fatal to beauty? For years strangeness has afflicted one of the basic concerns of physics: the nature of matter. When physicists considered matter on the smallest scale, it appeared to be an arbitrary jumble of elementary particles. No simple and orderly relationship among the particles could be perceived. Now at last the picture seems to be clearing up a bit. The very word "strangeness" has passed into the vocabulary of physics, but its share "in the proportion" is being reduced to a point where the beauty of order can be seen.

The new regularity can best be appreciated against the chaotic background from which it is emerging. To begin we should go back some 30 years to one of the most triumphant periods in the history of science. The theory of the atom stood essentially complete: nearly all the properties of ordinary matter could be mathematically deduced in terms of the motions of negatively charged electrons around positively charged nuclei. Most of the problems with which physics and chemistry had grappled during the preceding centuries were in principle solved. But at that time physicists began seriously to probe the interior of the atomic nucleus.

Then their troubles began. They soon learned that the nucleus is made up of protons and neutrons, but they could not explain nuclear properties only in terms of these constituents. Indeed, we still do not know exactly what their motions are. Furthermore it turned out that when a nucleus is shattered, entirely new types of matter are created—a bewildering variety of short-lived particles which apparently do not exist within the atoms of ordinary material. Some of them were reasonably well accounted for when they turned up, but others fit nowhere in the physicist's scheme of nature. They were called "strange" particles.

The First Particles

We are getting ahead of our story. We should begin in the early 1930s, when the atomic drama had only four characters: the electron, proton, neutron and photon. The first three are the building blocks of atoms—protons and neutrons in the nucleus and electrons in the space around it. The photon is the quantum unit of radiation; *i.e.*, it is the building block of the electromagnetic field.

The photon always travels with the velocity of light (denoted by the letter "c"); it can never be at rest. Because of its motion it possesses energy. It therefore also possesses mass, according to the famous relation $E = mc^2$. But the mass exists only by virtue of the motion. The electron, proton and neutron, on the contrary, can be at rest. Each has a mass when at rest and a corresponding rest energy. (When in motion, of course, they have additional energy and mass.)

The electron is the lightest particle with any rest mass, and this mass is a basic unit in subatomic physics. The size of the electron's negative charge is likewise a basic unit of electricity. In these units the proton has a mass of about 1,836.1 and a charge of plus one; the neutron has a mass of about 1,838.6 and no charge. The photon, as we have said, has no rest mass; also it has no charge, although it is the carrier of electromagnetic energy.

All these particles spin on their axes and, if they are charged, the spin makes them tiny magnets. According to the rules of quantum theory the spin has a fixed rate characteristic of the particle. In the system of units used in quantum theory, the characteristic spin of the electron, proton and neutron is 1/2; the spin of the photon is 1.

There is a further limitation on the spinning motion of these particles. If they are magnets, they are affected by external magnetic fields. In quantum mechanics the spin axis of each particle can assume only a few fixed directions with respect to an outside field. A particle with spin 1/2 can have two positions: its axis can point with or against the field. A particle with spin 1 can have

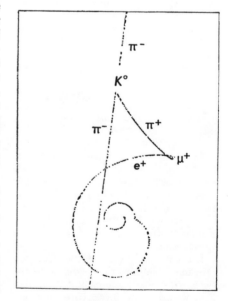

NEUTRAL K PARTICLE, formed in a collision of a negative pion with a proton, decays into a pair of oppositely charged pions, as outlined in the drawing. The

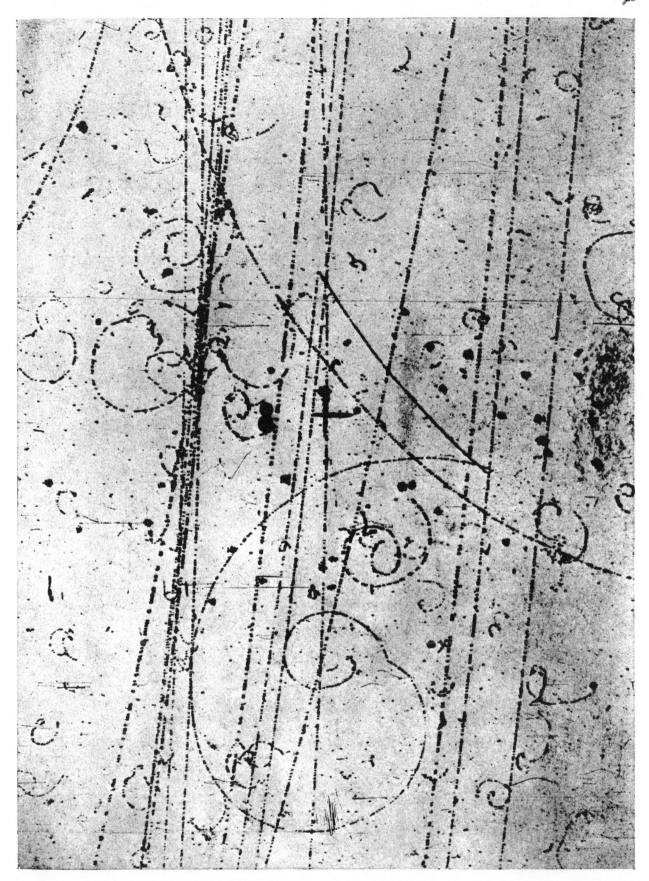

event is recorded in this photograph as a series of bubbles produced by charged particles in a chamber of liquid propane. The
positive pion decays further to a muon and a neutrino, which
leaves no bubble track. Finally the muon decays to a positron and
two invisible neutrinos. A neutral lambda was presumably made
together with the K, but if so it left the chamber without decaying
to charged particles and made no track. The incoming pion was
produced in the Cosmotron at Brookhaven National Laboratory.

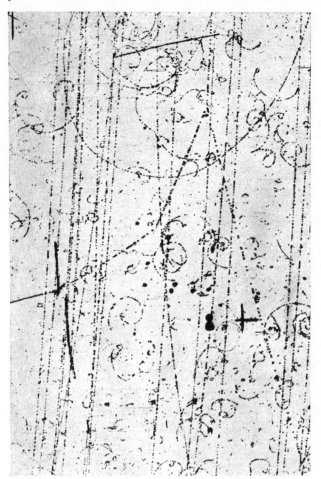
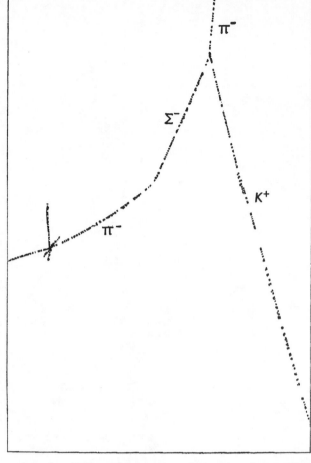

SIGMA AND K PARTICLES are produced together when a pion hits a proton in the bubble chamber. The sigma decays to a pion and an invisible neutron. The pion then hits a carbon nucleus and makes a "star." The experiments shown here and on the preceding page were performed by J. H. Steinberger, N. P. Samios, R. J. Plano, F. R. Eisler and M. Schwartz, all of Columbia University.

three positions: its axis can go with the field, perpendicular to or against the field [*see diagram on page 29*].

Another important property of particles, related to spin, is their "statistics." Electrons, protons and neutrons (and all other particles of spin 1/2) obey the famous exclusion principle. This says that only one particle of a kind can occupy a given quantum "state." Thus there can be only one electron at a time spinning in a particular direction and revolving in a given orbit around a nucleus. Particles which obey the exclusion principle are said to have Fermi-Dirac statistics: they are accordingly called fermions. Particles like the photon (and all other particles whose spins are whole numbers) do not obey the exclusion principle. They have Bose-Einstein statistics and are called bosons.

Interactions

So far we have been talking mainly about isolated particles. However, as will become increasingly evident, all particles are "coupled" to one another: when they are close together, they interact in various ways. The first such coupling to be recognized and studied was the one between the electron and the photon. It is this relationship which underlies quantum electrodynamics, the crowning achievement of atomic theory. Many physicists had a hand in the development of this theory, notably P. A. M. Dirac of England, Werner Heisenberg of Germany and Wolfgang Pauli, now in Switzerland. The theory explained the behavior of electrons in electromagnetic fields by saying that each electron continuously emits and absorbs photons. This pulsation is, so to speak, a "vital process" of the electron, and it is the means by which field and electron exert a force on each other.

We should point out that what has just been said does hardly more than name the theory. In quantum mechanics a theory is a set of mathematical relations which, given the interacting particles and the couplings between them, predicts their behavior in detail by yielding the probability of every possible reaction among the particles. Sometimes, particularly when the couplings are very strong, the mathematics turns out to be too difficult. Then the theory is not much help. In quantum electrodynamics, however, the mathematics is tractable, and this beautiful theory has successfully predicted the outcome of every fundamental atomic experiment at least as accurately as physical measurements can be made.

The basic "reaction" of quantum electrodynamics, in which electrons emit and absorb photons, is an example of what is called a virtual process. This concept, which concerns all the elementary particles, is peculiar to quantum theory. It involves an apparent violation of the law of the conservation of energy. The point is that a photon has energy; thus when a photon is spontaneously emitted by an electron, it would appear that the total energy of the system has suddenly increased. Quantum theory answers, in essence, that the photon is emitted and reabsorbed so fast that the gain in energy cannot be detected, even in principle. That is what is meant by a virtual pro-

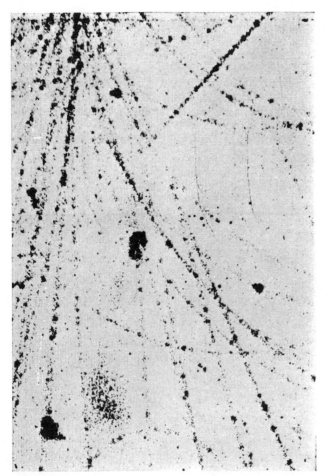

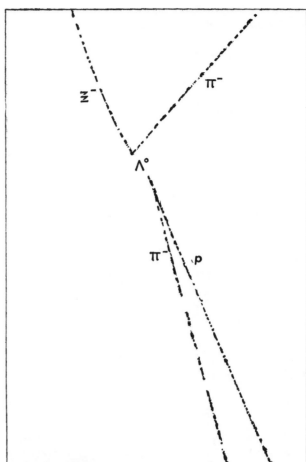

NEGATIVE XI PARTICLE decays into a neutral lambda and a negative pion. The lambda then decays into a proton and a second negative pion. The xi was produced by the collision of a high-energy cosmic ray with a nucleus in a lead plate. This photograph, which shows tracks in a cloud chamber below the plate, was made by E. W. Cowan of the California Institute of Technology.

cess. If the photon is undetectable, the conservation of energy is effectively not violated because, according to quantum mechanics, laws deal only with observable quantities. By adding enough energy from the outside (for example, by accelerating the electron) the photons can be converted from virtual to real particles.

Virtual photons are involved in every interaction between charged bodies and electromagnetic fields. The positive proton is also considered to emit and absorb virtual photons. Here, however, the theory is not quite so successful; its predictions for protons are not as accurate as those for electrons.

In an important sense the scheme we have described so far was complete and satisfactory. Between them the electron and photon sufficed to explain all the external properties of atoms; the proton and neutron accounted for the observed charges of atomic nuclei and roughly for their masses. There was, to be sure, nothing in the theories to explain why nature had chosen just these particles as her elementary building

blocks; but given that she had, they came close to being all that was needed.

Antiparticles

They did not come quite close enough. First of all, Dirac's theory of the electron predicted some additional particles [*see column II of chart on next two pages*]. It is well known that according to quantum theory a fundamental particle also has the properties of a wave. When Dirac's wave equation for the electron was solved, it yielded a negative frequency as well as a positive. Since frequency in quantum mechanics is proportional to energy, it was at first hard to see what the negative answer could mean. Dirac was able to prove that it does have a physical significance, and that it corresponds to an electron with positive charge. Furthermore, according to the theory, if a positive electron collided with a negative electron, they would annihilate each other and their mass would be converted into photons with an equivalent amount of energy. Conversely, if enough

energy could be concentrated in a small volume, as in a high-speed collision between two particles, a positive and a negative electron could be created.

These remarkable predictions were not actually made (although they were implied by the theory) until Carl D. Anderson of the California Institute of Technology discovered the positron. It had the mass of an electron and a unit of positive charge; when it met with a negative electron, the two were annihilated; it could be created, together with a negative electron, in energetic collisions. The positron is called the antiparticle of the electron because it cancels out an ordinary electron.

There are similar equations for the proton and neutron, so they also have their antiparticles. These have only been detected during the past two years [see "The Antiproton," by Emilio Segrè and Clyde E. Wiegand; SCIENTIFIC AMERICAN, Offprint 244]. Even the photon has an antiparticle in a mathematical sense. Here, however, the two solutions to the equation can be interpreted in the same way and the photon and antiphoton are indistinguishable. To put it another way, the photon is its own antiparticle.

The Neutrino

The second necessary addition to the list of particles arose out of the behavior of the neutron. Inside the nucleus a neutron can live indefinitely. But when the particle is observed outside,

it proves to be unstable. In an average time of about 18 minutes it spontaneously ejects a beta particle (the same thing as an electron) and turns into a proton. The proton and electron together are about 1.5 electron masses lighter than the neutron, so this amount of mass appears to be lost in the decay; it is equivalent to some 780,000 electron volts of energy. This should show up as

the kinetic energy of the decay products, but in fact the proton and electron rarely have so much energy. To account for the discrepancy Pauli suggested that another particle, with zero rest mass and almost undetectable, also is formed in the decay, and that it carries off the missing energy. Enrico Fermi, who pursued the idea, named the invisible particle the neutrino. Reasoning

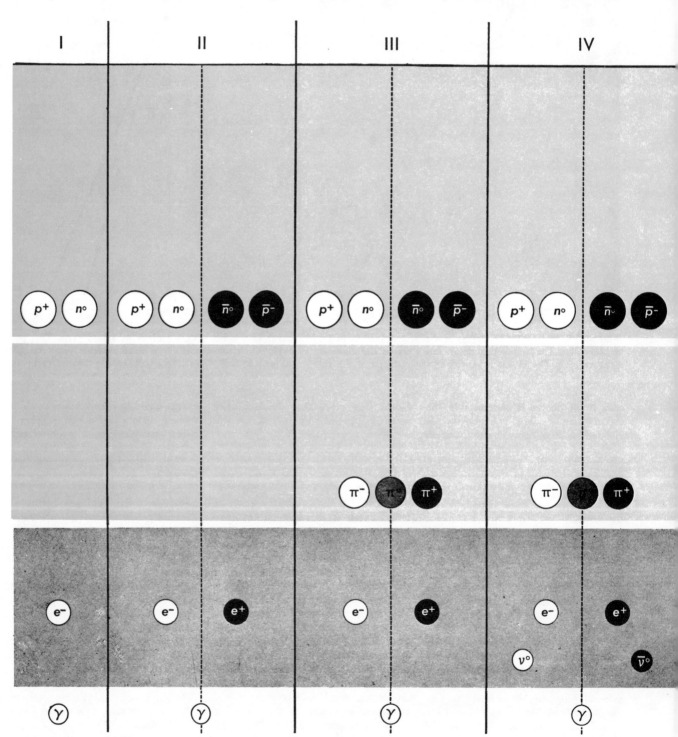

TABLE OF PARTICLES traces their increase over the past 25 years. Particles added in columns II-IV were first predicted theoretically. Those added in V and VI were discovered by experiment. "Ordinary" particles are shown as white balls, antiparticles as black balls. The neutral pion and the photon is each its own antiparticle. The top group comprises the heavy particles and the next lower group the

by direct analogy with Dirac's process for electrons and photons, Fermi constructed a complete theory of beta-decay. Its fundamental process is that a neutron continuously loses and regains an electron and a neutrino by virtual emission and absorption. (Strictly speaking the "neutrino" involved is actually the antineutrino.) Although Fermi's reaction was written as a virtual process, the emission or decay process can become real without the addition of outside energy because the mass lost in the decay provides the energy needed.

The Pion

The last particle to be added to the list was predicted by another analogy with the Dirac process. The problem was to describe the force that holds protons and neutrons (which may jointly be called nucleons) together in the nucleus. Since electromagnetic forces had been successfully explained in terms of the photon or field quantum, it was logical to try the same approach with nuclear forces. The Japanese physicist Hideki Yukawa took this step. He proposed that nucleons emit and absorb a

mesons. The color over both shows they are all strongly coupled. The wide band between the groups signifies that heavy particles are conserved. The third group, in gray, comprises the light particles.

A bar over a symbol stands for an antiparticle, but the convention is not used for the pion, muon or electron. For explanation of K_1^0 and K_2^0 see page 36 of the text.

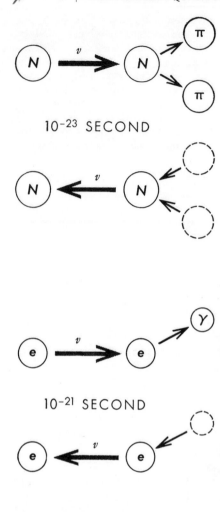

10⁻²³ SECOND

10⁻²¹ SECOND

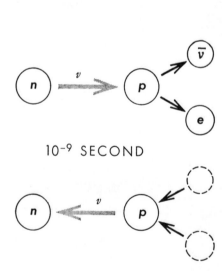

10⁻⁹ SECOND

BASIC PROCESSES are shown schematically. In the strong Yukawa interaction (top) a nucleon "virtually" emits pions (above) and absorbs them (below). Time scale for the process is listed between the two reactions. In the electromagnetic interaction (center) an electron (or other charged particle) virtually emits and absorbs a photon. In the weak and very much slower Fermi interaction (bottom) a neutron virtually emits and absorbs an electron and antineutrino.

nuclear-field quantum called a meson, just as electrons emit and absorb photons. From the known properties of the nuclear force Yukawa was able to deduce some of the characteristics of the meson. The fact that the force extends only over a very short range could be shown to mean that the meson, unlike the massless photon, would have a finite rest mass. There were also various reasons to suppose that there would be both charged and neutral mesons.

Yukawa's conjecture was fully confirmed, but only after more than 10 years. The particle he predicted has been found and is now called the pi meson, or pion. It weighs about 270 electron masses, and comes in three forms: positive, negative and neutral [see "Pions," by Robert E. Marshak; SCIENTIFIC AMERICAN, Offprint 226].

The emission of pions by a nucleon must of course be virtual, since the pions possess energy including energy in the form of rest mass. According to the theory the strength of the nuclear force field should depend on the number of quanta outside the emitting particle. The nuclear force is so strong that a nucleon must emit pions very frequently; there must usually be more than one outside the nucleon at the same time. In fact, the current conception of protons and neutrons is that they consist of some sort of core surrounded by a pulsating cloud of pions. As in the case of photons, if enough energy is supplied pions will materialize into real particles. Since the pion's mass is equivalent to 135 million electron volts (mev) of energy, it requires at least this amount to make one real pion.

Here we come to the end of the first part of the story. We have already accumulated a rather large number of "elementary" particles, but at least they all seem to make sense. In fact, most of them were predicted theoretically before they were actually discovered. As has already been pointed out, the neutrino has an antiparticle. The negative pion is the antiparticle of the positive pion, and vice versa; the neutral pion, like the photon, is its own antiparticle. Finally real pions, like the neutron, are unstable. After a very short time they decay into other particles [see table on page 30].

Twelve Particles

We might call the ideas we have sketched so far the dozen-particle theory of matter [see column IV in the chart on the preceding two pages]. As we have said, it is a fine theory for explaining the

properties of atoms. It is rather crude in its attempt to account for the inner workings of the nucleus, but it does explain them in a general way. And in any event it makes a good case for each of the particles. They all have an explicit role to play and they emerge naturally from the theory.

What is more, the 12 fall into four well-defined groups: (1) heavy particles, consisting of the nucleons (proton and neutron) and their antiparticles; (2) mesons, or particles of intermediate weight; (3) light particles, consisting of the electron and neutrino and their antiparticles; and (4), in a class of its own, the photon. We may also note that the heavy and light particles, which are the "ordinary" constituents of matter, have spin 1/2 and are fermions [see table on page 30]. The mesons and the photon, which are field quanta, have spin zero and are bosons. The groups are interconnected by three basic reactions. The Yukawa process connects heavy particles with mesons, the Dirac process connects light particles with photons, and the Fermi process connects heavy particles with light particles [see chart on this page].

Of course the particles also behave according to the more general laws of physics. They obey the conservation of energy, and of linear and angular momentum. They also obey the conservation of charge. So far as we know the net amount of electric charge in the universe never changes. When charged particles are created out of energy, they can be created only in particle-antiparticle pairs, with each new positive charge offset by a negative. And in every particle reaction the net charge of the bodies entering the reaction must equal that of the products.

Another conservation principle arises out of the evident stability of nuclear matter. All the experimental evidence indicates that it is never created or destroyed; that is to say, that the number of nucleons must remain constant. Thus a proton can be created out of energy, but only together with an antiproton. The two cancel each other both mathematically and, when they come together, physically.

Particle Reactions

We should mention two other characteristics of particle reactions which appear to operate as general laws. First, the reactions are reversible. If one particle is observed to split into two others, we expect to find that the pair can also combine to form the original particle.

Second, the emission of a particle is related to the absorption of the corresponding antiparticle: if we know the "cross section" or probability for one, we can compute the probability for the other. For example, consider the beta-decay process [see reactions 1-3 in the table on page 32]. Since a neutron turns into a proton by emitting an electron and an antineutrino we expect to find the reverse reaction in which a proton absorbs an electron and an antineutrino, turning into a neutron [reaction 2]. Furthermore, since the absorption of an electron corresponds to the emission of an antielectron, or positron, we should also have the reaction in which a proton absorbs an antineutrino and emits a positron, again turning into a neutron [reaction 3]. This reaction is, in fact, the one by which the neutrino was finally detected experimentally.

Thus the rules provide a kind of algebra with which we can "solve" problems in particle physics. Let us see how they work for a sample problem—the decay of the pion. The neutral pion is found experimentally to decay into a pair of photons in a very short time: about 10^{-15} seconds. We want to show why it should do this [reactions 4-8]. We start out with the basic reaction involving pions, the Yukawa reaction in which a nucleon, let us say a proton, emits a virtual pion. We are interested in what the pion will do, so we should like to have an equation that starts out with a neutral pion followed by an arrow. By reversing the equation and transposing terms (changing particles into antiparticles) we arrive at the desired equation [reaction 6], which tells us that the pion turns (virtually) into a proton-antiproton pair. Such a pair annihilates, and can yield photons. Thus we have arrived at the desired result: a neutral pion decaying into two photons.

To someone seeing it for the first time, this chain of reasoning may seem just a trivial shuffling of symbols. But each shuffling summarizes a detailed (and often difficult) calculation of probabilities. Thus in the end we arrive at a reasonably exact prediction of what will happen, how long it should take, and so on. On the other hand, our knowledge of the situation is so incomplete that an apparently airtight chain of reasoning can also result in completely wrong answers.

The Muon

There is no better illustration of this than the decay of the charged pion.

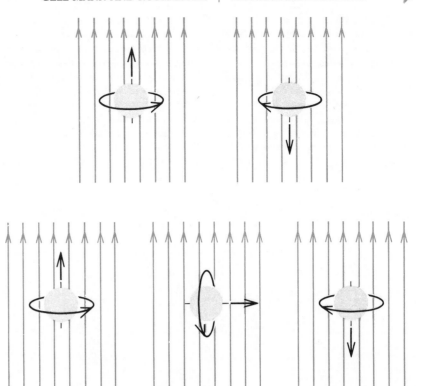

SPINNING PARTICLE can take only certain fixed positions with respect to an outside magnetic field (colored arrows). Particles with spin 1/2 (top) can align their axes with or against the field. Particles with spin 1 (bottom) can go with, across or against the field.

Using our basic reactions and rules we can "prove" that the positive pion should decay to a positron and a neutrino [reactions 9-14]. But if the reaction ever happens, it is so rare that it has never been observed. How does the positive pion actually decay? It yields a neutrino and a totally new particle: the muon!

Here we have nature at her most perverse. She has given us a particle for which there is no theoretical justification and no use whatever. The muon was the unwelcome baby on the doorstep, signifying the end of days of innocence. The situation was further complicated by an unfortunate historical accident which for a long time made it impossible even to classify the muon. The muon was detected before the pion, and everyone took it to be the meson Yukawa had predicted. For this role, however, its properties are all wrong. It notably does not interact strongly with nucleons and thus could not be the particle responsible for the nuclear-force field. Thus until the pion was discovered the muon made even less sense than it does today.

Now that we know what the muon is not, we can at least accept it for what it is. It comes with both positive and negative charge (the negative pion decays into a negative muon and anti-

neutrino). It weighs about 207 electron masses and has a spin of 1/2 (i.e., it is a fermion). It lives for about a millionth of a second and then decays into an electron, a neutrino and an antineutrino. The positive muon must of course yield a positive electron, or positron, and the negative muon a negative electron. Each muon is the other's antiparticle.

Although the muon does not come out of the dozen-particle theory—indeed, it demonstrates that such a theory is incomplete if not wrong—it can be connected to the other particles. To see the connection we must re-examine our fundamental processes—the Dirac or electromagnetic interaction, the Yukawa or nuclear interaction and the Fermi or beta-decay process. It turns out that these processes differ enormously in "strength." The Yukawa process is known as a strong interaction; it accounts for the great force that holds nucleons together in the nucleus. Electromagnetic forces are some 137 times weaker than nuclear forces. Another indication of strength is the probability that a process will occur in a given time; i.e., its average rate. The strong interactions are as fast as anything can possibly be. The emission or absorption of a pion takes place in some 10^{-23} seconds, which is just about the time it would take a light ray to cover a distance equal to the diameter of a nucleon. The electromagnetic process is of course 137 times slower.

What about the Fermi interaction? It is incomparably weaker than the others. The factor is about 10^{-14}—it is a hundred thousand billion times weaker than the strong interactions! Furthermore, all the processes involving neutrinos—beta-decay and the decay of pions and muons—are about equally weak. Thus the muon participates in one of the three fundamental types of interaction. (As a charged particle, of course, it participates in the electromagnetic interaction as well.) Also, since it is a light fermion, it seems to group itself naturally with the electron and neutrino.

A glance at the table of lifetimes on this page will show that the decays which we have said are equally weak have widely different times. But speed is supposed to be an indication of strength. The answer is that speed is not determined solely by strength. It also depends on the energy available to make the reaction go. In the case of the neutron, which takes an average of 18 minutes to decay, there is very little available energy; the difference between the mass of the decaying particle and the mass of its decay products is only slightly more than the mass of one electron. For pion and muon decays there is much more energy; they are correspondingly faster. If corrections are made in each case for the available energy, it turns out that a kind of "intrinsic" speed for all the weak processes is very close to 10^{-9} seconds, which is 10^{14} times slower than the strong interaction.

It is surely remarkable that all the weak processes have the same strength, and it is probably significant. Nature is trying hard to tell us something, but so far we have been unable to decipher the message.

Strange Particles

With the muon nature gently warned physicists that they had not yet divined her innermost secrets. Then around 1950 she rudely introduced a whole procession of new particles. They were utterly unexpected and had properties which could not be explained on the basis of previous theory.

The new arrivals showed up first in the showers of particles which occur when high-energy cosmic rays strike a lead plate inside a cloud chamber. Among the tracks of the showers were found some curious two-pronged or V-shaped patterns that could not be explained by any known particle process. Physicists were forced to conclude that some unknown neutral particle (which would leave no track in the cloud chamber) had decayed into two charged particles. The neutral particle presumably had been made in the lead plate. Once people started looking for the so-called V-particles, they turned out to be very common.

As V-events were collected and studied, it became clear that there were at least two new neutral particles. One, which decays into a proton and a negative pion, was named the lambda; the other, which decays into a positive and a negative pion, was called the K.

When they had recovered from the shock, physicists began to try to fit the new particles somehow into the general scheme. From the pattern of its decay (into a fermion and a boson) the lambda

PARTICLE		SPIN	REST MASS (ELECTRON MASSES)	MEAN LIFE (SECONDS)	DECAY PRODUCTS
XI	Ξ^-	$\frac{1}{2}$	2585	10^{-10} TO 10^{-9}	$\Lambda^\circ + \pi^-$
	Ξ°	$\frac{1}{2}$	NOT YET FOUND		
SIGMA	Σ^+	$\frac{1}{2}$	2325	$.7 \times 10^{-10}$	$p + \pi^\circ$ $n + \pi^+$
	Σ^-	$\frac{1}{2}$	2341	1.5×10^{-10}	$n + \pi^-$
	Σ°	$\frac{1}{2}$	2324	NOT MEASURED	$\Lambda^\circ + \gamma$
LAMBDA	Λ°	$\frac{1}{2}$	2182	2.7×10^{-10}	$p + \pi^-$ $n + \pi^\circ$
PROTON	p	$\frac{1}{2}$	1836.1	STABLE	
NEUTRON	n	$\frac{1}{2}$	1838.6	ABOUT 1,000	$p + e^- + \bar{\nu}$
K MESON	K^+	0	966.5	1.2×10^{-8}	$\mu^+ + \nu$ $\pi^+ + \pi^\circ$ $\pi^+ + \pi^+ + \pi^-$ $\pi^+ + \pi^\circ + \pi^\circ$ $\mu^+ + \nu + \pi^\circ$ $e^+ + \nu + \pi^\circ$
	K^-	0	966.5	1.2×10^{-8}	$\mu^- + \bar{\nu}$ $\pi^- + \pi^\circ$ $\pi^- + \pi^- + \pi^+$ $\pi^- + \pi^\circ + \pi^\circ$ $\mu^- + \bar{\nu} + \pi^\circ$ $e^- + \bar{\nu} + \pi^\circ$
	K_1°	0	965	1×10^{-10}	$\pi^+ + \pi^-$ $\pi^\circ + \pi^\circ$
	K_2°	0	965	3×10^{-8} TO 10^{-6}	$\pi^+ + e^- + \bar{\nu}$ $\pi^- + e^+ + \nu$ $\pi^+ + \mu^- + \bar{\nu}$ $\pi^- + \mu^+ + \nu$ $\pi^+ + \pi^- + \pi^\circ$ $\pi^\circ + \pi^\circ + \pi^\circ$
PION	π^+	0	273.2	2.6×10^{-8}	$\mu^+ + \nu$
	π^-	0	273.2	2.6×10^{-8}	$\mu^- + \bar{\nu}$
	π°	0	264.2	10^{-16} TO 10^{-15}	$\gamma + \gamma$
MUON	μ^-	$\frac{1}{2}$	206.7	2.2×10^{-6}	$e^- + \nu + \bar{\nu}$
ELECTRON	e^-	$\frac{1}{2}$	1	STABLE	
NEUTRINO	ν	$\frac{1}{2}$	0	STABLE	
PHOTON	γ	1	0	STABLE	

PROPERTIES OF PARTICLES are being determined with increasing precision. This table presents the latest experimental values. Spins of the K mesons and the strange heavy particles are still doubtful. For the unstable particles which have more than one mode of decay the table lists all the sets of products now known. Others may still be discovered.

can be shown to be a fermion, presumably with spin 1/2. It is subject to the law of conservation of nucleons. Since one nucleon is produced in the decay, one must have been used in the formation process. For example, the lambda might be made in a collision between a proton and a negative pion—the reverse of the decay process. (Some other particle, such as a neutral pion, would also have to be made to carry off the excess energy.) The frequency with which lambdas appear shows that they are made by a strong process. The mass of the lambda turns out to be 2,181 electron masses.

The K particle has to be a boson because it decays into two pions, both bosons. Its spin must then be a whole number (most likely zero). It cannot be made out of a nucleon, because there are no nucleons in its decay products. It is, however, produced frequently, and thus by a strong process. It has a mass of 965.

From these first considerations it is possible to classify the new particles to some extent. The lambda obviously belongs with the nucleons or heavy particles. It is made from a nucleon and, like them, it is a fermion. The K, on the other hand, is a boson; hence it is put in the meson group along with the pions.

The lambda and K particles were only the beginning. Soon other particles were identified. In the same category with the lambda are the sigma particles, charged and neutral, and the negative xi particle. Added to the neutral K were a pair of charged particles of about equal mass, called the positive K particle and the negative K particle.

Long Lifetimes

The very existence of all these distinct forms of matter is a difficult problem. But if we accept the fact that they exist, their behavior presents us with even deeper questions. The trouble arises with the decay of the new particles. Their lifetimes range from about 10^{-8} to 10^{-10} seconds, which is on the time scale of the weak interactions. But the particles are made, as we have seen, by strong interactions, the time scale of which is some 10^{-23} seconds. According to one of our most fundamental tenets—that of reversibility—a particle made in a strong interaction should also decay that way.

The new particles would seem to have ample opportunity to decay by strong processes. Consider, for example, the neutral lambda particle and let us play our equation-juggling game [see reactions 15-19 on next page]. By using two

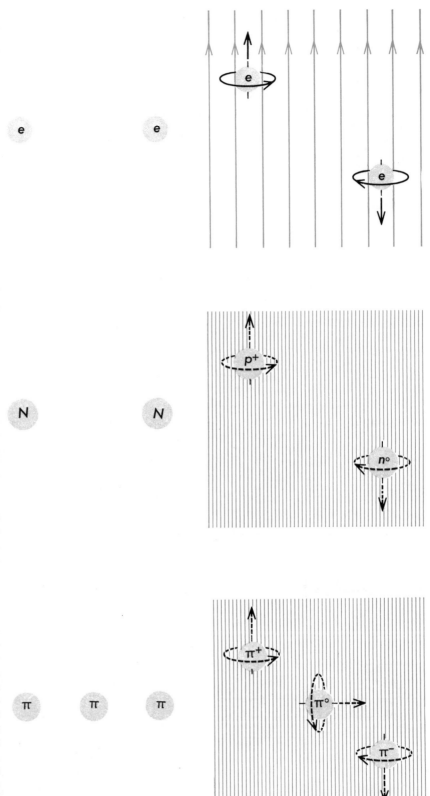

DIFFERENT STATES of a particle can be recognized only through certain interactions. Two oppositely spinning electrons (*top*) would seem alike in the absence of an external magnetic field (*left*). When the field, indicated by colored arrows, is turned on, the electrons separate into different energy states (*right*). Similarly, in the absence of the electromagnetic interactions, all nucleons (*center*) are indistinguishable, as are all pions (*bottom*). When these interactions, indicated by closely spaced colored lines, are taken into account, isotopic "spin," which is represented by the broken arrows, separates the nucleons into protons and neutrons, and separates the pions into their three different charge types.

PARTICLE REACTIONS

1.	$n \longrightarrow p + e^- + \bar{\nu}$	*Fermi process. Neutron splits into proton, electron and antineutrino.*
2.	$p + e^- + \bar{\nu} \longrightarrow n$	*Reverse reaction. Proton, electron and antineutrino combine into neutron.*
3.	$p + \bar{\nu} \longrightarrow e^+ + n$	*Transpose electron as antiparticle. Proton interacts with antineutrino to yield positron and neutron.*
4.	$p \overset{v}{\longrightarrow} p + \pi^\circ$	*Yukawa process. Proton splits (virtually) into proton and neutral pion.*
5.	$p + \pi^\circ \overset{v}{\longrightarrow} p$	*Reverse reaction. Proton and pion combine into proton.*
6.	$\pi^\circ \overset{v}{\longrightarrow} p + \bar{p}$	*Transpose proton as an antiproton. Pion splits into proton and antiproton.*
7.	$p + \bar{p} \longrightarrow \gamma + \gamma$	*Proton and antiproton annihilate to yield photons.*
8.	$\pi^\circ \longrightarrow \gamma + \gamma$	*Net result. Neutral pion decays to photons.*
9.	$p \overset{v}{\longrightarrow} n + \pi^+$	*Yukawa process. Proton splits (virtually) into neutron and positive pion.*
10.	$n + \pi^+ \longrightarrow p$	*Reverse reaction. Neutron and pion combine into proton.*
11.	$\pi^+ \overset{v}{\longrightarrow} p + \bar{n}$	*Transpose neutron as an antineutron. Pion splits into proton and antineutron.*
12.	$p + \bar{\nu} \longrightarrow e^+ + n$	*Fermi process. Proton interacts with antineutrino to yield positron and neutron.*
13.	$p + \bar{n} \longrightarrow e^+ + \nu$	*Transpose antineutrino and neutron as their antiparticles. Proton interacts with antineutron to yield positron and neutrino.*
14.	$\pi^+ \longrightarrow e^+ + \nu$	*Net result. Positive pion decays to positron and neutrino.*
15.	$\pi^- + p \longrightarrow \Lambda^\circ + \pi^\circ$	*Hypothetical process. Negative pion and proton interact to yield lambda and neutral pion.*
16.	$\Lambda^\circ + \pi^\circ \longrightarrow \pi^- + p$	*Reverse reaction. Neutral pion interacts with lambda to yield negative pion and proton.*
17.	$\Lambda^\circ \overset{v}{\longrightarrow} \pi^\circ + \pi^- + p$	*Transpose neutral pion (it is its own antiparticle). Lambda decays (virtually) into a neutral pion, a negative pion and a proton.*
18.	$p + \pi^\circ \longrightarrow n$	*Yukawa process. Proton absorbs neutral pion.*
19.	$\Lambda^\circ \longrightarrow p + \pi^-$	*Net result. Lambda decays quickly to proton and negative pion.*
20.	$\pi^- + p \longrightarrow \Lambda^\circ + K^\circ$	*Hypothetical process. Pion and proton interact to yield lambda and neutral K.*
21.	$\Lambda^\circ + K^\circ \longrightarrow \pi^- + p$	*Reverse reaction. Lambda and K interact to yield pion and proton.*
22.	$\Lambda^\circ \overset{v}{\longrightarrow} \pi^- + p + \bar{K}^\circ$	*Transpose K as antiparticle. Lambda decays (virtually) into pion, proton and anti-K.*
23.	$p + \pi^- \longrightarrow n$	*Yukawa process. Proton absorbs pion, turning into neutron.*
24.	$\Lambda^\circ \longrightarrow n + \bar{K}^\circ$	*Net result. Lambda decays into neutron and anti-K.*

strong processes, one known and one hypothetical but plausible, we arrive at the statement that the lambda is converted to a proton and a negative pion, which is its actual mode of decay. There is plenty of energy available for the process: the lambda's mass is 74 units greater than that of the proton and pion, which gives an energy difference of 37 mev. Thus we have "proved" that the lambda must decay as fast as it is made. The same thing can be demonstrated for all the other new particles. The only trouble is that they live 100,000 billion times longer than they should! It was this enormous discrepancy between their expected and observed lifetimes that was chiefly responsible for the designation "strange" or "queer" particles.

Associated Production

After contemplating the situation for a couple of years a number of theoreticians, in particular A. Pais of the Institute for Advanced Study, were able to suggest a possible resolution of the paradox. Their idea was that strange particles are made only in groups of two or more at a time. The concept is now known as associated production. It implies that the strong interaction which manufactures a strange particle somehow works only on more than one at a time. The trick here is that a strong process of this kind would not be reversible because of lack of energy.

For example, suppose that a lambda and a K were made in the collision of a negative pion with a proton. Now we apply our reaction rules to this process in order to predict the fate of the lambda [*reactions 20-24*]. We arrive at the conclusion that it "decays" into a proton and an anti-K. But of course this is impossible, because the two daughter particles have a combined mass greater than that of the parent. A thorough analysis shows that every possible case of associated production leads to a similar result for the separate decay of any one of the strange particles that are made. The possible avenues of decay always turn out to require too much energy. Thus, by moving away from each other immediately after they are created, the strange particles are saved

PARTICLE EQUATIONS summarize certain detailed chains of reasoning discussed in the text. Some of the chains of reasoning are found to be faulty, as indicated by the fact that the reactions they predict do *not* take place. These have been shown in gray.

from death by strong interaction, and they live until the much less probable weak processes catch up with them.

At first there was little or no experimental evidence for associated production. However, when the Cosmotron at the Brookhaven National Laboratory began to make strange particles on order, it appeared that the rule was indeed obeyed. In fact, the very first reaction to be discovered was the pion-proton collision suggested above, leading to the associated production of the lambda and the neutral K particles.

Now the question arose: What does associated production mean? Can it be related to any other principles? It tells us that strong reactions involving single strange particles are forbidden. When nature rules out an event, her legislation often takes the form of a conservation law. Such-and-such cannot happen because something must be conserved. To take a simple example, we never see a particle decay into products whose total mass is greater than its own. The conservation of energy forbids it.

Once the rule of associated production had been found, it was natural to ask whether there might not be a conservation law behind it. If the law could be discovered, we might find out much more about strange particles. Associated production says they must be made more than one at a time. But are all combinations possible or are some ruled out? The conservation law should tell.

Isotopic Spin

It appears that the law has been discovered, and we do know some of the rules by which strange particles are made. In order to see what the principle and the rules are, we must go back to the older particles and to a concept which we have not mentioned until now. This is the notion of "isotopic spin."

First let us have another look at ordinary spin. Imagine that we have a pair of isolated electrons which we can actually see as small specks. So far as we can tell, they are identical. We believe they are spinning, but they are so small that we cannot detect their motion. Now we put them in a magnetic field. Obeying the laws of quantum mechanics, their spins line up with or against the field. Suppose one goes with and the other against. Then the two particles have different energies, and we can distinguish one from the other. This imaginary experiment underlines the fact that the electron is a "doublet" so far as its magnetism is concerned. It may be in one of two possible energy

states. But without an external magnetic field there is no way to tell the states apart; they "degenerate" into indistinguishability.

Now in the early days of modern nuclear physics—soon after the discovery of the neutron—a situation arose that was reminiscent of this magnetic "degeneracy." Experiments on the deflection of moving protons and neutrons by other protons and neutrons disclosed the surprising fact that the nuclear force, or strong interaction, between nucleons is the same regardless of the type of particle involved. The forces between two protons, two neutrons, or a proton and a neutron are all equal. This phenomenon, called charge independence, means that so far as strong interactions are concerned the neutron and proton look like the same particle. They can be distinguished only by their electromagnetic interaction. Suppose electromagnetism could be "turned off" like a magnetic field in the laboratory. Then the proton and neutron would also degenerate into indistinguishability. Hence the nucleon can be thought of as a "charge doublet," with one state representing the proton and the other the neutron.

This idea occurred to Heisenberg, who proceeded to express it mathematically. He constructed a mathematical description of the nucleon which included a variable that could take on just two values. One value thus represents the proton and the other the neutron. The mathematics is very much like that used by Pauli to describe the spin of an electron. Thus Heisenberg called his quantity isotopic spin. "Isotopic" refers to the fact that in a sense the proton and neutron are isotopes: they have nearly the same mass but different charge. "Spin" is purely an analogy, and a somewhat misleading one at that. It simply reflects a similarity to the mathematical term for real spin.

Isotopic spin, then, is a mathematical device which distinguishes the proton and neutron; physically they are distinguished by their different couplings to the electromagnetic field. The analogy to real electron spin is very close: the isotopic spin of the nucleon is also 1/2. Like real spin, it has components $+1/2$ and $-1/2$ with respect to a given, or reference, direction. In quantum electrodynamics it is customary to place a particle in a system of coordinates, the "Z" axis of which is parallel to the surrounding magnetic field. Hence the reference direction for isotopic spin is also considered to go along the Z axis, and the components are denoted by I_z. The convention has been adopted that I_z of $+1/2$

represents the proton and I_z of $-1/2$ the neutron. In Heisenberg's mathematical theory charge independence becomes a conservation law. When nucleons interact, the total isotopic spin is conserved. This statement can be shown to mean the force is equal between a proton and a proton, a neutron and a neutron, and a proton and a neutron. So far, it should be emphasized, the idea of isotopic spin is a mere formality. It adds nothing to the notion of charge independence, and is simply another convenient way of expressing it mathematically.

Charge Multiplets

When Yukawa explained nuclear forces in terms of pion emission and absorption, isotopic spin took on a somewhat broader significance. The British physicist Nicholas Kemmer, now at Edinburgh University, realized that the concept must also apply to pions. His reasoning was as follows: Nuclear forces, involving the virtual exchange of pions, are charge-independent. Therefore pions must be charge-independent, and the isotopic spin concept should apply.

Now we recall that the pion has three possible charges: plus, minus and neutral. Thus it constitutes a charge "triplet" which, if charge could be turned off, would also degenerate into indistinguishability. In the case of real spin, a triplet means that the particle has a spin of 1, since it may then assume three different directions with respect to the field. Its Z components are $+1$, 0 and -1. So we assign an isotopic spin of 1 to the pion, and we say that its components with respect to the reference direction (its I_z's) are $+1$, 0 and -1.

The grouping of particles into charge doublets or triplets (collectively called multiplets) provides a convenient short-

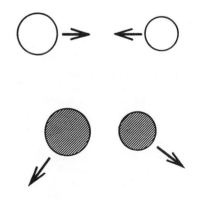

ASSOCIATED PRODUCTION is suggested in this diagram. Two normal particles collide (*top*), making a pair of strange particles (*bottom*) which immediately separate.

hand way of identifying them. If we say that the pion is a triplet with its center of charge at zero [see chart on opposite page], this tells us at once that the charges are $+1$, 0 and -1, the isotopic spin is 1, and the I_z's are $+1$, 0 and -1. Similarly, to say that the nucleon is a doublet with center of charge at $+1/2$ means that the charges are 0 and $+1$, the isotopic spin $1/2$ and the I_z's $+1/2$ and $-1/2$. As can be seen in the chart, the antinucleon is another doublet centered at $-1/2$. Its isotopic spin is $1/2$ and its I_z is $-1/2$ and $+1/2$.

Note that the isotopic spin and charge multiplet concepts give us another difference between nucleons and pions. Nucleons are a doublet centered at $+1/2$, while the pions are a triplet centered at 0.

Now let us turn to the strange particles. An obvious question is whether their interactions are also charge-independent and conserve isotopic spin. Are positive sigma particles, for example, just like negative and neutral ones except for their electric properties? There is no direct experimental evidence on the point, but it seemed reasonable to suppose that charge independence would apply to the strong interactions of the new particles as it had been found to apply to the strong Yukawa coupling. This would mean that the strange particles are charge multiplets. If so, it was generally supposed that they would follow the same classification as the nucleon and pion. That is, the heavy strange particles seem to be related to nucleons; they are made out of nucleons and decay back to nucleons. Therefore the heavy particles were generally thought to be doublets, having an isotopic spin of $1/2$ and a charge center at plus or minus $1/2$. The K particles, on the other hand, apparently belong with the pion, so it was supposed they would fall into a triplet, with an isotopic spin of 1 and charge center at 0.

About five years ago one of the authors of this article (Gell-Mann) and the Japanese physicist Kazuhiko Nishijima independently conceived the idea that the strange particles might not follow this arrangement. Furthermore, the departure from the expected arrangement might account for their strange behavior. In the case of the present author it was a matter of discovery by slip of the tongue. Discussing the heavy strange particles one day, he spoke of them as having an isotopic spin of 1, but then quickly corrected himself, saying "I mean a half, of course."

The more he thought about the "mistake" later on, the more he began to wonder whether it really was one. How do we know that heavy particles are doublets with isotopic spin $1/2$? To be sure, the particles seemed to be related to the nucleon—and for the sake of order and simplicity one certainly hoped that they were related. But if they were members of that family, they were strange members. Perhaps it was precisely in their isotopic spin that their strangeness lay. Suppose the heavy particles, instead of being doublets of isotopic spin $1/2$, like the nucleon, were triplets of isotopic spin $+1$ or even singlets of isotopic spin 0. (A particle with zero isotopic spin has only one possible state and is thus a singlet.) Suppose the K particles, instead of being triplets like the pion, were doublets? (At the time the table of strange particles was just being filled in by experiment. It was not even known, for example, whether there were charged K or lambda particles.)

After toying with the idea for a while, the author began to see that it might contain in it just the conservation law that was needed to explain associated production and the strangely long lifetimes of strange particles. In a moment we shall try to show roughly how this comes about. First let us pursue the idea a little further.

Displaced Multiplets

Recall that a simple way of describing a group of particles is to indicate its center of charge and whether it is a doublet or triplet. The nucleon is a doublet with center at $+1/2$; the pion is a triplet with center at 0, and so on. Now suppose that among the heavy particles there is a singlet at charge 0 [see chart on opposite page]. Could this, by any chance, be the neutral lambda? If it is, note that the center of this "multiplet" (a singlet is a multiplet with one member) is at 0—one-half charge unit less than the center of the nucleon doublet. The original expectation was that all heavy particles would have their multiplet centers at $+1/2$. Therefore the lambda is "displaced" by $-1/2$ charge unit. Perhaps this displacement is the essential physical characteristic of the particle which accounts for its "strangeness." Let us assume that it is; in fact, let us invent a new physical quantity and call it strangeness. For reasons of mathematical convenience we make the strangeness equal to twice the displacement. Thus the strangeness of our putative lambda particle is twice the displacement of $-1/2$, or -1. (The strangeness of the nucleon is of course 0. Its

charge center sets the reference point from which the displacements of other heavy particles are measured.)

Next we observe that, in our system of classification by multiplets, the antinucleons form a doublet which is an image of the nucleon doublet, mirrored on the zero-charge line [see chart]. Thus the other heavy particles should also have antiparticles in corresponding multiplets. We accordingly place in our table an antilambda, which is also at 0. Its displacement is $+1/2$ (from the "normal" charge center of the antinucleon). Hence its strangeness is $+1$.

Now we can try something else—say a triplet centered at 0. Its strangeness would be -1. If there is such a triplet, there should be three strange particles, positive, negative and neutral, all with approximately the same mass. At the time the strangeness theory was conceived, no such triplet was known. Now it has apparently been found in the sigma particle (Σ^+, Σ°, Σ^-). Again we expect a corresponding multiplet of antiparticles.

Still another possibility is a heavy-particle doublet displaced a full charge unit, from $+1/2$ to $-1/2$, i.e., with strangeness -2. This would mean a pair of particles with charge -1 and 0. We now believe that the negative member of the pair is the xi particle (Ξ^-). The neutral member (Ξ°) has not yet been detected, but the general success of the strangeness theory gives us considerable confidence that it will turn up.

The K particles may fall into doublets like the nucleon and antinucleon. This would mean that the K^+ and K° constitute one doublet with its charge center at $+1/2$. Since these particles are grouped with the pion, whose "natural" charge center is at zero, their displacement is $+1/2$ and their strangeness $+1$. The K^- then is part of a doublet together with a second neutral K, the anti-K°, which is the antiparticle of the first. The

STRANGENESS is illustrated in tabular form. Particles (white circles) and antiparticles (black circles) are grouped in multiplets with their charges indicated by the colored vertical lines. The solid colored carets mark the charge center of each multiplet; open carets mark the "expected" location of charge centers ($1/2$ for heavy particles, $-1/2$ for heavy antiparticles and 0 for mesons). Horizontal colored arrows show the displacement of each center from the expected position. The strangeness equals twice the value of this displacement.

PARTICLE	ISOTOPIC SPIN	STRANGENESS	CHARGE				
			−1	−½	0	+½	+1
NUCLEON	$\frac{1}{2}$	0			$n°$		p^+
ANTI-NUCLEON	$\frac{1}{2}$	0	$\bar{p}^-$		$\bar{n}°$		
LAMBDA	0	−1			$\Lambda°$		
ANTI-LAMBDA	0	+1			$\bar{\Lambda}°$		
SIGMA	1	−1	Σ^-		$\Sigma°$		Σ^+
ANTI-SIGMA	1	+1	$\bar{\Sigma}^-$		$\bar{\Sigma}°$		$\bar{\Sigma}^+$
XI	$\frac{1}{2}$	−2	Ξ^-		$\Xi°$		
ANTI-XI	$\frac{1}{2}$	+2			$\bar{\Xi}°$		$\bar{\Xi}^+$
PION	1	0	π^-		$\pi°$		π^+
K	$\frac{1}{2}$	+1		$K_1°$ $K_2°$	$K°$		K^+
ANTI-K	$\frac{1}{2}$	−1	$\bar{K}^-$		$\bar{K}°$	$K_1°$ $K_2°$	

center of this doublet is −1/2, its displacement is −1/2 and its strangeness is −1. The pion plays the same role in the meson group as the nucleon does in the group of the heavy particles. That is, its charge center provides the reference point for measuring strangeness, so its own strangeness is 0.

Conservation of Strangeness

Now we have assigned a strangeness to all the strongly coupled particles. What is the point of this exercise? Simply this. It is possible to prove from the principle of charge independence that strangeness must be conserved in the strong and electromagnetic interactions. That is to say, in any reaction of these two types the total strangeness of the particles entering the reaction must equal the total strangeness of the products. We will show that this conservation law accounts for the observed behavior of strange particles.

To begin with, it obviously "explains" associated production. Strange particles are made in collisions between ordinary particles. The strangeness of the latter is 0. Therefore the total strangeness of the products must be 0. This means that at least two must be made at a time so that their individual strangenesses offset each other. Consider the case we have already mentioned: the production of a lambda and a neutral K from the collision of a pion and a proton. The lambda has a strangeness −1 and the K° has strangeness +1: total strangeness, 0.

As we have already seen, associated production explains why strange particles do not decay by strong interaction. But they must also be immune to the electromagnetic process, since their lifetimes are on the time scale of the weak interactions. The law of conservation of strangeness shows us how the particles avoid decay by electromagnetism as well as by strong interactions.

We can indicate only crudely how the law does so. It can be shown that the conservation of strangeness is mathematically equivalent to the conservation of the Z component of isotopic spin: I_z. The latter quantity is essentially a measure of charge: in any multiplet, the greater the I_z, the greater the charge. (For example, in the nucleon doublet I_z's of −1/2 and +1/2 correspond to charges of 0 and +1; in the pion triplet I_z's of −1, 0 and +1 correspond to charges of −1, 0 and +1, etc.) Electromagnetic interactions are thought to depend only on charge. By a general rule of quantum mechanics this means that they should conserve I_z (which meas-

ures charge). But to say they conserve I_z is the same as saying they conserve strangeness. Hence an isolated particle whose strangeness is not 0 cannot decay into particles with zero strangeness by an electromagnetic process.

Eventually, of course, the strange particles do decay into ordinary ones, and on a time scale about the same as that for the weak interactions. So it seems that the strange-particle disintegrations are members of this great class of processes. Weak interactions, therefore, do not conserve strangeness. It has recently been discovered that they also violate another conservation law, the conservation of parity, which has to do with symmetry in nature between right and left [see "The Overthrow of Parity," by Philip Morrison; SCIENTIFIC AMERICAN, April, 1957]. Whether there is any deep connection between the two laws and their violation is not yet known. In any case, it is clear that nature has been concealing some of her most important secrets in the weak processes, and that one of the major jobs facing physics today is to discover the laws which govern these processes.

Selecting Particles

When we began the search for a conservation law to account for associated production, it was with the hope that it would tell us still more about the birth and death of strange particles. And so it does. For instance, the rule of associated production would permit a reaction in which two neutrons collide to form a pair of lambdas.

In fact, this was considered one of the more likely reactions. But the reaction has never been observed, and the conservation of strangeness tells us that, practically speaking, it never will be observed. The neutron's strangeness is 0 and the lambda's is −1. Thus the neutron collision, if it yields a lambda, must also yield another particle with a strangeness of +1, such as a neutral K: for example $(n+n\rightarrow\Lambda^\circ+n+K^\circ)$.

Once again let us consider the case of the sigma and K particles. The sigma is a triplet with strangeness −1. The K particles are a pair of doublets; the pair including the K^+ has a strangeness of +1, while the pair including the K^- has strangeness −1. Hence it is possible to make a Σ^- (strangeness −1) and a K^+ (strangeness +1) together, but not a Σ^+ and a K^-, both of which have strangeness −1. The first reaction has been discovered in pion collisions with protons $(\pi^-+p\rightarrow\Sigma^-+K^+)$. Aside from the conservation of strangeness, there seems no

reason why the reaction should not also go the other way $(\pi^-+p\rightarrow\Sigma^++K^-)$. But this has not been found to happen, and the strangeness principle tells us why it has not.

As a still further example of the power of the strangeness rule, we may examine the decay of the neutral sigma. The sigma triplet has the same strangeness (−1) as the lambda. The sigma is also some 150 electron masses heavier. Therefore it should be possible, both from the standpoint of available energy and of conservation of strangeness, for the sigma to decay to the lambda. That is to say, the sigma should not have to wait for the weak processes to end its life. However, in the case of a charged sigma, there would have to be some other charged particle in the decay products to conserve charge. It might be a pion $(\Sigma^+\rightarrow\Lambda^\circ+\pi^+)$. The pion's strangeness is 0, so the strangeness accounts are in balance. But the pion's mass is 270, or about 120 more than is available energetically. The neutral sigma, on the other hand, does not need to produce any other charged particles. Its excess energy can be carried away by photons $(\Sigma^\circ\rightarrow\Lambda^\circ+\gamma)$. This reaction has in fact been observed. It is an electromagnetic process (since it involves photons) and is therefore only a little slower than the strong interactions themselves.

Thus strangeness gives us rules for selecting the possible strange particles and their possible decays. As a matter of fact, a few particles were predicted by the strangeness table before they were actually found. The only one still missing is the neutral xi.

The Neutral K

Before leaving this "periodic table" of strange particles, one final comment is in order. It will be noticed that the K° and its antiparticle are also listed as a different pair of particles called the K°_1 and K°_2. One of the most striking successes of the strangeness theory was the prediction of this situation. The reasoning which led to the prediction is too complicated to set forth here, but it indicates a remarkable shuffling process on the part of nature. The K° and anti-K° are made in different processes. Once made, each of them can decay in two different ways, one of which takes a little longer than the other. Quantum theory shows that only half of each type of particle can follow either mode of decay. Thus we have two different manufacturing processes and two different decay processes, with a reshuffling in between. Nature segregates the neutral K particles on one basis in their manufacture and on another in their decay. She makes $K°$'s and anti-$K°$'s. After they are made, half of each of these particles "become" $K°_1$'s and half become $K°_2$'s. This is demonstrated by the way they decay.

In the strangeness theory, then, we have a means of classifying strange particles. The theory is consistent with the fundamental idea of four groups of particles and three types of reaction. Thus we still have only the heavy particles (some of them strange), the mesons (some of them strange), light particles and the photon. And the couplings between them are strong, electromagnetic or weak.

At present our level of understanding is about that of Mendeleyev, who discovered only that certain regularities in the properties of the elements existed. What we aim for is the kind of understanding achieved by Pauli, whose exclusion principle showed *why* these regularities were there, and by the inventors of quantum mechanics, who made possible exact and detailed predictions about atomic systems.

We should like to know the laws of motion of the particles; to predict, among other things, how they will interact when they collide and how these interactions will deflect one particle when it collides with another. As this article is written a number of physicists are hard at work on theories which they hope may supply the laws. Time will be the judge.

On a still more fundamental level there are questions to which the answers seem as yet much more remote. Are all the particles we have mentioned really elementary, or are some of them just compounds of other particles? If so, which are elementary and which are not? Why has nature chosen to use this particular set of particles to build the material world? Why are the charges of elementary particles limited to the values $+1$, -1 and 0? These and many other such puzzles seem to lie entirely beyond the power of our present theories. Shall we ever know the answers? Every physicist has an abiding faith that we shall. But it will probably require some wholly new ideas. For one thing, many theoreticians believe that the present concepts may be entirely inapplicable at extremely short distances —of the order of the dimensions of the particles. In fact, it is suspected that here these concepts become self-contradictory.

It is likely to be quite a while before the particle physicist finds himself out of a job.

4

Strongly Interacting Particles

by Geoffrey F. Chew, Murray Gell-Mann and Arthur H. Rosenfeld
February, 1964

Presenting an account of recent developments in high-energy physics. These particles that respond to the strongest of the four natural forces no longer seem "elementary." They may be composites of one another

Only five years ago it was possible to draw up a tidy list of 30 subatomic particles that could be called, without too many misgivings, elementary. Since then another 60 or 70 subatomic objects have been discovered, and it has become obvious that the adjective "elementary" cannot be applied to all of them. For this reason the adjective has been carefully avoided in the title of this article. There is now a widespread belief among physicists that none of the particles with which this article is mainly concerned deserves to be singled out as elementary.

What is happening has happened before in physics: the old way of looking at things, which was adequate for perceiving order in a limited number of observations, finally proved cumbersome and inadequate when the accuracy and range of observation increased. This happened with the Ptolemaic scheme of epicycles for describing the motions of the planets. Much the same thing occurred early in this century when spectroscopists, studying the light emitted by excited atoms, found a profusion of discrete wavelengths that were at total variance with the wavelengths predicted by classical electrodynamics. The spectroscopists accumulated so much empirical information, including sets of "selection rules" governing the permissible states of excited atoms, that it finally became possible in 1926 for Werner Heisenberg, Erwin Schrödinger and others to formulate a new mechanics—quantum mechanics—capable of predicting most of the states of matter on the atomic and molecular scale.

A similar situation may exist today in particle physics. The great unifying invention analogous to quantum mechanics is still not clearly in sight, but the experimental data are beginning to fall into striking and partly predictable patterns. What can be said to summarize the vast amount of particle information now available?

First of all, there is a clear distinction between strongly interacting particles, such as the neutron and proton, and other particles. The neutron and proton are known to interact through the strong, short-range nuclear force, which is responsible for the binding of these particles in atomic nuclei. All particles discovered to date participate in this strong interaction except the photon (the particle of light and other electromagnetic radiation) and the four particles called leptons: the electron, the muon (or mu particle) and the two kinds of neutrino.

Another striking property of the strongly interacting particles is that none of them has a small rest mass. Rest mass is the mass that a particle would have if it were motionless; this is the minimum mass the particle can have. It is now common to express this mass as its equivalent in energy, rather than in units of the electron's mass, as was often done in the past. The lightest strongly interacting particle is the pion (or pi meson), which has a mass with an energy equivalent of some 137 million electron volts (Mev). In contrast, the mass of the electron is about .5 Mev and that of the photon and the neutrinos is believed to be zero.

A third general observation is that the recent proliferation of particles has so far occurred almost exclusively among the strongly interacting particles. Although this proliferation came as a surprise to physicists, a precedent for this state of affairs can be found in ordinary atomic nuclei. It is well known that all compound nuclei, from the nucleus of deuterium (heavy hydrogen) to those of the heaviest elements, can exist at a variety of energy levels, comprising a "ground" state and many excited states.

These levels, which can be detected in several ways, indicate different degrees of binding energy among the component nucleons (neutrons and protons) in the nucleus. The binding energy, of course, is an expression of the strong nuclear force.

It is now clear that the nuclear force can similarly give rise to numerous states among those strongly interacting particles sometimes designated elementary. The lower states are "bound," or stable; the higher states are only partly bound, or unstable, decaying in a tiny fraction of a second. The result is that all strongly interacting particles exhibit a spectrum of energy levels with no sharp upper limit.

Since the leptons do not participate in strong interactions, it is not surprising that their spectrum of states, beginning with the massless neutrino and apparently terminating sharply at the muon, with a mass of 106 Mev, bears no resemblance to any known dynamical spectrum. In recent years physicists have learned much about the simplicity and regularity in the properties of leptons, but they have learned nothing of why these particles exist.

In the following discussion we shall begin by considering the place of the strong force in the hierarchy of four forces that seem to underlie all the operations of the physical universe. Next we shall describe a new nomenclature that assigns each of the strongly interacting particles to one of a small number of families, each characterized by a distinctive set of properties. One group of these families embraces the baryons, which in general are the heaviest particles; a second group consists of mesons, the first members of which to be discovered were lighter than the baryons. The new naming system will require a brief review of the seven quantum num-

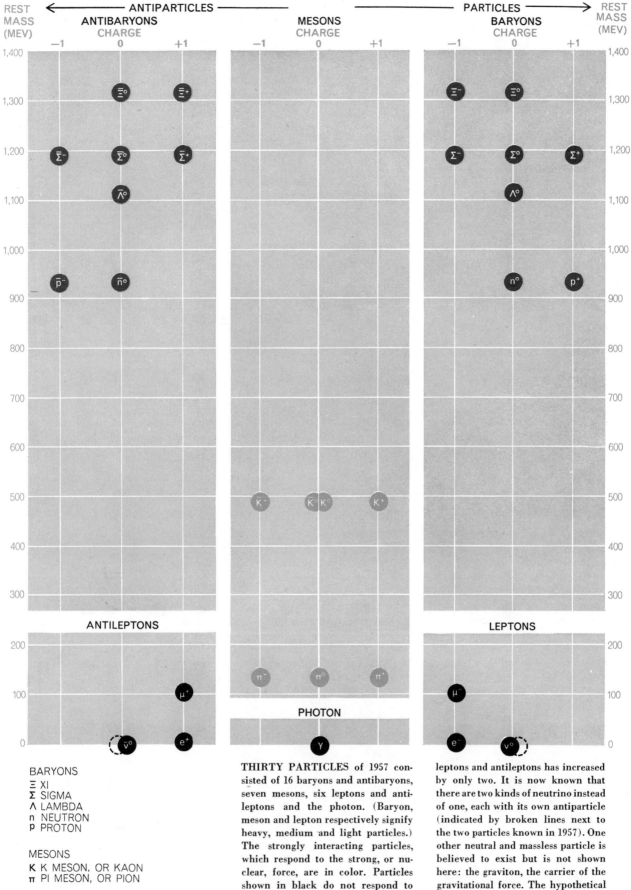

REST MASS (MEV)

← ANTIPARTICLES →

ANTIBARYONS
CHARGE
−1 0 +1

MESONS
CHARGE
−1 0 +1

← PARTICLES →

BARYONS
CHARGE
−1 0 +1

REST MASS (MEV)

ANTILEPTONS

LEPTONS

PHOTON

BARYONS
Ξ XI
Σ SIGMA
Λ LAMBDA
n NEUTRON
p PROTON

MESONS
K K MESON, OR KAON
π PI MESON, OR PION

LEPTONS
μ MU, OR MUON
e ELECTRON
ν NEUTRINO

THIRTY PARTICLES of 1957 consisted of 16 baryons and antibaryons, seven mesons, six leptons and antileptons and the photon. (Baryon, meson and lepton respectively signify heavy, medium and light particles.) The strongly interacting particles, which respond to the strong, or nuclear, force, are in color. Particles shown in black do not respond to this force. It is the former that have proliferated in the past half-dozen years, as shown on the next two pages. In the same period the number of leptons and antileptons has increased by only two. It is now known that there are two kinds of neutrino instead of one, each with its own antiparticle (indicated by broken lines next to the two particles known in 1957). One other neutral and massless particle is believed to exist but is not shown here: the graviton, the carrier of the gravitational force. The hypothetical carrier of the weak force, also not shown, should have a considerable mass and one unit of electric charge. Evidence for it is now being sought.

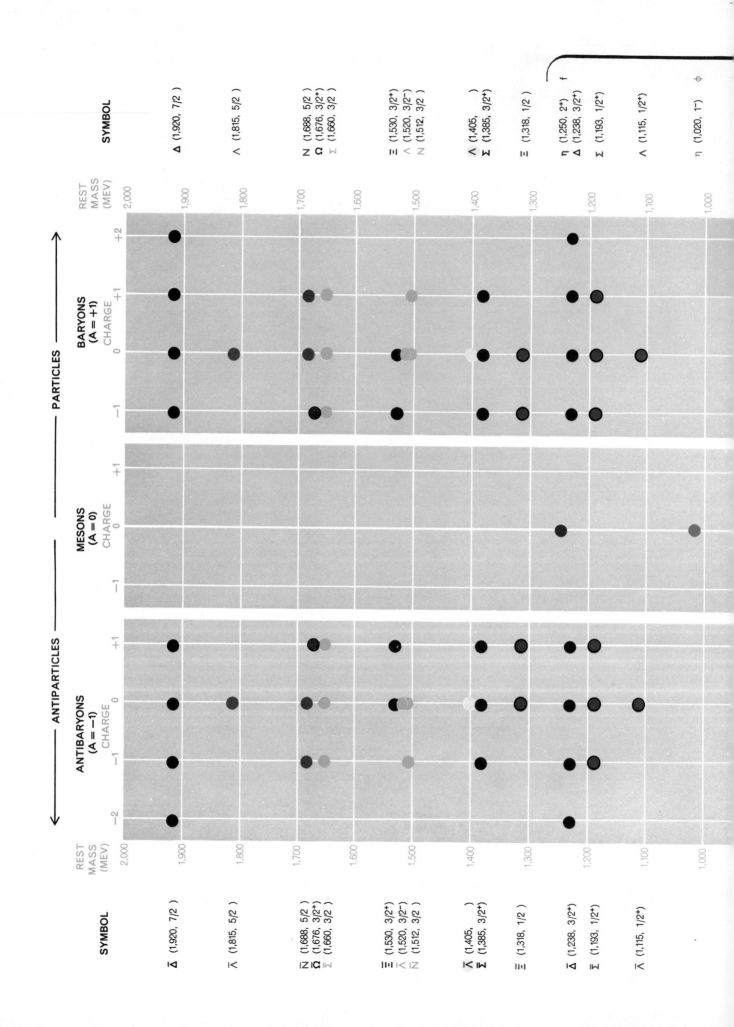

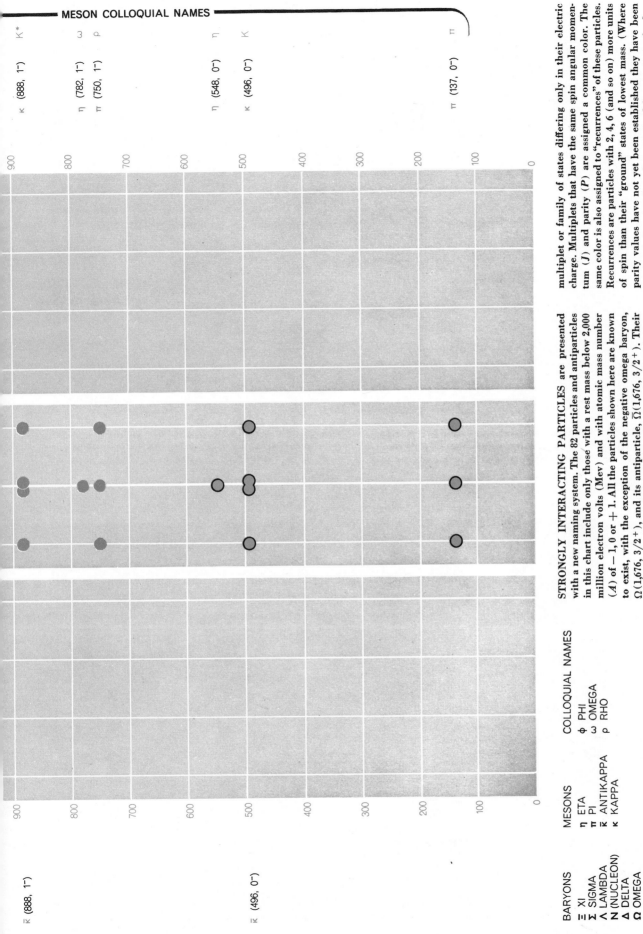

K*   K* (888, 1⁻)

ω   ω (782, 1⁻)
ρ   ρ (750, 1⁻)

η   η (548, 0⁻)
K   κ (496, 0⁻)

π   π (137, 0⁻)

900 800 700 600 500 400 300 200 100 0

STRONGLY INTERACTING PARTICLES are presented with a new naming system. The 82 particles and antiparticles in this chart include only those with a rest mass below 2,000 million electron volts (Mev) and with atomic mass number (A) of -1, 0 or $+1$. All the particles shown here are known to exist, with the exception of the negative omega baryon, $\Omega(1{,}676, 3/2^+)$, and its antiparticle, $\bar{\Omega}(1{,}676, 3/2^+)$. Their existence has been predicted by the "eightfold way." The numbers in parentheses are explained at lower left. The mass assignments are averages for the members of a charge multiplet or family of states differing only in their electric charge. Multiplets that have the same spin angular momentum (J) and parity (P) are assigned a common color. The same color is also assigned to "recurrences" of these particles. Recurrences are particles with 2, 4, 6 (and so on) more units of spin than their "ground" states of lowest mass. (Where parity values have not yet been established they have been left blank, but they have been guessed for purposes of color-coding.) Stable and metastable particles are circled in black; unstable particles (also called resonances) are uncircled.

BARYONS
Ξ XI
Σ SIGMA
Λ LAMBDA
N (NUCLEON)
Δ DELTA
Ω OMEGA

MESONS
η ETA
π PI
K̄ ANTIKAPPA
κ KAPPA

COLLOQUIAL NAMES
φ PHI
ω OMEGA
ρ RHO

LETTER A = ATOMIC MASS, OR BARYON, NUMBER
PARENTHESES CONTAIN: (REST MASS IN MILLION ELECTRON VOLTS, SPIN ANGULAR MOMENTUM, PARITY)

K̄ (888, 1⁻)

K̄ (496, 0⁻)

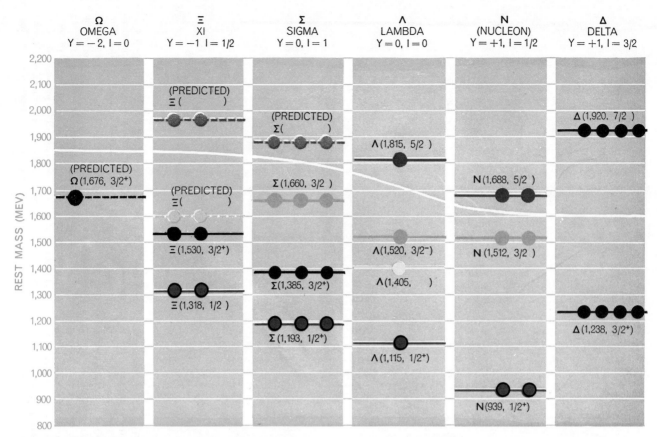

BARYON MULTIPLETS that have the same values of hypercharge (*Y*) and isotopic spin (*I*) are arranged in columns. Only six combinations of *Y* and *I* are known at present, each identified by an uppercase Greek letter. The particle symbols and color code are those presented on the preceding two pages. Particles above the break in the tan background are recurrences of low-lying states.

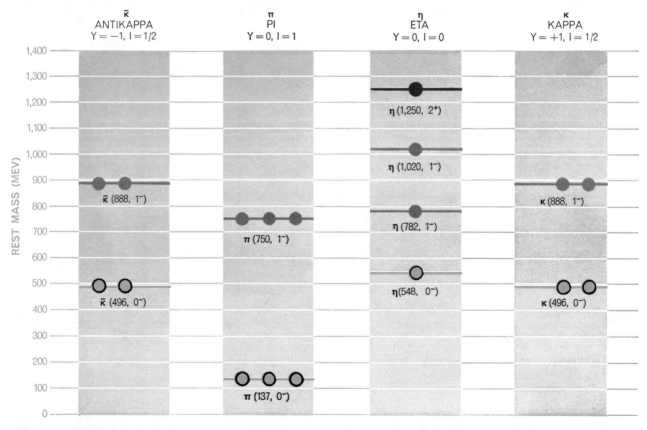

MESON MULTIPLETS with like values of *Y* and *I* are arranged in columns. Four combinations of *Y* and *I* are identified by lower-case Greek letters. No meson recurrences have yet been identified. The first predicted recurrence is a pi triplet at about 1,700 Mev.

bers, or physical quantities, that are conserved in strong interactions.

We shall next describe how these quantities are conserved when particles decay into different "channels," representing different modes of decay. This will lead to a description of "resonances," or unstable particles, which account for most of the proliferation among strongly interacting particles.

We shall then be ready to discuss two classification systems, or rules for the formation of groups, that have brought to light deep-seated family relations among strongly interacting particles. These rules have made it possible to predict the existence of still undiscovered particles, their approximate masses and certain other properties. One system is based on the concept of the "Regge trajectory"; the other is the "eightfold way."

Next we shall explain why the term "elementary" has fallen into disrepute for describing strongly interacting particles. There is growing evidence that all such particles can be regarded as composite structures. Finally we shall describe the "bootstrap" hypothesis, which may make it possible to explain mathematically the existence and properties of the strongly interacting particles. According to this hypothesis all these particles are dynamical structures in the sense that they represent a delicate balance of forces; indeed, they owe their existence to the same forces through which they mutually interact. In this view the neutron and proton would not be in any sense fundamental, as was formerly thought, but would merely be two low-lying states of strongly interacting matter, enjoying a status no different from that of the more recently discovered baryons and mesons and the nuclei of atoms heavier than those of hydrogen.

Forces and Reaction Times

In present-day physics the concepts of force and interaction are used interchangeably. The strong, or nuclear, force is the most powerful of the four basic interactions that, together with cosmology, account for all known natural phenomena. (Cosmology provides the stage on which the forces play their roles.) The strong interaction is limited to a short range: about 10^{-13} centimeter, which is about the diameter of a strongly interacting particle.

The next force, in order of strength, is the electromagnetic force, which is about 1 per cent as powerful as the

strong force. Its strength decreases as the square of the distance between interacting particles, and its range is in principle unlimited. This force acts on all particles with an electric charge and involves the uncharged photon, which is the carrier of the electromagnetic-force field. The electromagnetic force binds electrons to the positively charged nucleus to form atoms, binds the atoms together to form molecules and thus in its manifold workings is responsible for all chemistry and biology.

Next in order, with only about a one-hundred-trillionth (10^{-14}) of the strength of the strong interaction, is the weak interaction. It also has a short range and cannot, as far as anyone knows, bind anything, but it governs the decay of many strongly interacting particles and is responsible for the decay of certain radioactive nuclei. It is most easily studied in the behavior of the four leptons, which do not respond to the strong force.

The fourth and weakest force is gravity, which has only about 10^{-39} of the strength of the strong interaction. It produces large-scale effects because it is always attractive and operates at long range. On the scale of atomic nuclei, however, its effects are undetectable.

Many particles are "coupled" to all four of these interactions. Take, for example, the proton. It is a strongly interacting particle, and since it is elec-

trically charged it must also "feel" the electromagnetic force. It can be created by the beta decay of a neutron, a decay in which the neutron emits a negative electron and an antineutrino by a weak-interaction process; hence it must be involved in weak interactions. And like all other matter the proton is attracted by gravity. The least reactive particle is the neutrino, which is directly coupled only to the weak interaction and to gravitation. The neutrino shares with the other leptons a total immunity to the strong force.

An important idea, not self-evident in the foregoing, is that the basic forces can do more than bind particles together. For instance, when two particles collide and go off in different directions (the phenomenon called scattering), an interaction is involved. If a particle is moving with enough energy before striking a particle at rest, a new particle can be created in the collision. The collision of a proton and a neutron can yield a proton, a neutron and a neutral pion, or it can yield two neutrons and a positive pion. The collision can also yield strongly interacting particles that are more massive than either of the colliding particles. This, in fact, is the process by which particle-accelerating machines have created the scores of new particles heavier than protons and neutrons. Thus the basic forces are

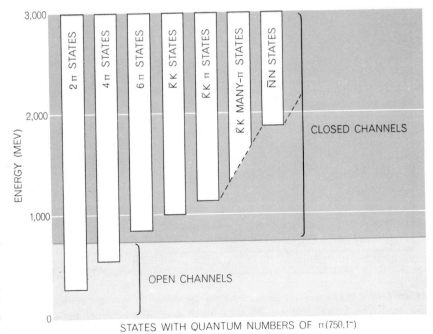

"COMMUNICATING CHANNELS" (*vertical bars*) are nuclear states that possess all the same quantum numbers as a particular particle, in this case $\pi(750, 1^-)$. Channels into which it has sufficient energy to decay are "open" channels. Channels for which it lacks sufficient energy are "closed." The threshold energy for gaining access to a communicating channel is the sum of the rest masses (in Mev) of the various particles constituting the channel.

interactions that can scatter, create, annihilate and transform particles.

The interactions of principal interest in high-energy physics take place when one of the particles in the interaction is traveling at nearly the speed of light, or more than 10^{10} centimeters per second. Since the size of a particle is typically about 10^{-13} centimeter, the minimum reaction time is less than 10^{-23}

second for a particle moving at the speed of light. What we mean when we call the strong interaction "strong" is that even in that brief time the strong force is powerful enough to cause a reaction to take place. Electromagnetic reactions, being 100 times weaker than strong reactions, take around 100 times longer, or typically 10^{-21} second. Processes involving the weak interaction,

which is 10^{-14} times weaker than the strong interaction, commonly take about 10^{-9} second.

Conservation Laws

When one of the present authors (Gell-Mann) and E. P. Rosenbaum discussed particles in these pages not quite seven years ago [see "Elementary Particles," SCIENTIFIC AMERICAN, Offprint 213], 30 well-established particles and antiparticles were singled out for attention. In this collection there were 16 baryons and antibaryons, seven mesons, six leptons and antileptons and the photon [see illustration on page 39]. At that time it was customary to classify as elementary not only the photon and the leptons but also all the baryons and mesons—the strongly interacting particles.

A distinction, which we now regard as unjustified, was drawn between these strongly interacting particles and the states of ordinary atomic nuclei containing two or more nucleons, which, of course, are also strongly interacting. These nuclei, such as the deuteron (the nucleus of heavy hydrogen) and the alpha particle (the nucleus of helium), had been classified as composite structures, made essentially of protons and neutrons, almost from the beginning of nuclear physics because of the small binding energies involved in them. This article will place little emphasis on such nuclei. We shall concentrate on the lighter particles, not because we believe them to be more elementary than their heavy brothers but because their status is still in doubt. If they are in fact composite dynamical structures, their binding energies are often enormous. Furthermore, if elementary particles do exist, they will certainly not include the obviously composite nuclei.

The chart on pages 40 and 41, which omits the photon and leptons, shows 82 particles and antiparticles, all of them strongly interacting, and the list has been arbitrarily limited to baryons and mesons with a rest mass less than 2,000 Mev. Most of the 82 particles belong to family groups that have acquired pet names comparable to, but usually less elegant than, the names given to those in the 1957 list. It would be unreasonable to expect the reader to master this specialized vocabulary, which reflects chiefly the confused state of high-energy physics a few years ago. We shall therefore introduce the reader to a new system of nomenclature that has developed quite recently and that provides a great deal of information about each particle. Although it may look forbidding

RESONANT CAVITY analogy accounts for the appearance of unstable particles called resonances. The "cavity" for the $\pi(750, 1^-)$ meson is shown at top with an energy source present (left) and removed (right). Energy (colored arrows) can flow into the cavity through one channel and leave through one or more open channels. The 6π channel, however, is closed because access requires a higher frequency (that is, more energy) than $\pi(750, 1^-)$ possesses. The sinusoidal curves show that at certain frequencies the cavity resonates, and when the energy is turned off, the resonance can persist for several cycles. The first resonance corresponds to $\pi(750, 1^-)$ itself. A second resonance with the same quantum numbers could conceivably exist. The colored curve at right represents the amplitude of the resonant waves when the source amplitude is kept constant and the frequency is varied.

at first sight, it is really no harder to master than the telephone company's all-digit dialing system.

The new classification scheme takes advantage of the fact that nature conserves many quantities (in addition to energy and momentum) and shows various symmetries (such as that between left and right). As a result groups of particles have similar properties, which, as we shall see, can be indicated by a common notation. There is a close relation between symmetries and conservation laws, and in a particular case one can refer either to a symmetry or to the associated conservation law, whichever is more convenient. The conserved quantity appears in quantum mechanics as a quantum number, which is often either an integer (such as 0, 1, 2, 3 and so on) or a half-integer (such as 1/2, 3/2, 5/2 and so on).

Some conservation laws appear to be universal: they are obeyed by all four basic interactions. This inviolable group includes the conservation of energy, of momentum, of angular momentum (the momentum associated with rotation) and of electric charge. Another exact conservation law is best described as a kind of mirror-image symmetry. It is the symmetry between particles and antiparticles, in which whatever is left-handed for one is right-handed for the other. For each particle there is an antiparticle with the same mass and lifetime but with some properties, such as electric charge, reversed. Some neutral particles, such as the photon and the neutral pion, are their own antiparticles.

In the new system for naming strongly interacting particles we shall make use of five quantities, each indicated by a letter symbol, that are conserved by the strong interactions but not necessarily by the electromagnetic or weak interactions. These five quantities are: atomic mass number (A), hypercharge (Y), isotopic spin (I), spin angular momentum (J) and parity (P). The chart on the next page should help the reader to keep these five quantum numbers in mind as we discuss them in more detail. Also included in the chart are two other quantities that are conserved by strong interactions but that are not essential to the naming system: electric charge (Q) and a quantity called G that has only two values, +1 and −1, and can be assigned only to mesons that have a hypercharge of 0.

The first three quantum numbers—A, Y and I—provide the basis of the naming system. What these three numbers do, in effect, is to describe the geometric pattern of the particles as they are arranged in the chart on pages 40 and

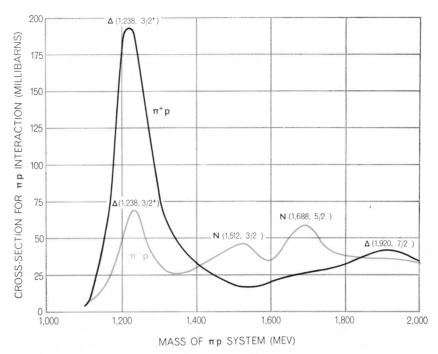

FIRST RESONANCE, the unstable particle called Δ(1,238, 3/2+), was discovered by Enrico Fermi and his colleagues in 1952. The resonance appears when protons are bombarded with high-energy pions. When the interaction "cross section" is plotted against the effective mass of the pion-proton system, a peak is found at around 1,238 Mev. The peak is much larger for the π+p interaction than for the π−p interaction. Other resonances occur at 1,512, 1,688 and 1,920 Mev, each peak corresponding to an unstable particle.

41. There it will be seen that mesons and baryons occur in "charge multiplets," or families of states differing only in their electric charge. The number of particles and their charges occur in different patterns: singlets, doublets, triplets and quadruplets. Only 10 different patterns are known or predicted at present, and each pattern represents a different set of values for A, Y and I. As we shall explain, each of the 10 patterns is identified by a different Greek letter.

Now we shall describe the physical significance of A, Y, I, J and P, but for convenience we shall discuss them in a slightly different order to emphasize certain relations among them. A is simply the long-familiar atomic mass number used to describe atomic nuclei. It is also known as baryon number. Like electric charge, A can be 0, ± 1, ± 2, ± 3 and so on. For uranium 235, A is 235, indicating that the nucleus of this isotope contains 235 neutrons and protons, for each of which A equals 1. Neutrons and protons are baryons and, by definition, so are all other particles with an A of 1. Particles with an A of − 1 are antibaryons. Mesons have an A of 0 (as do the leptons and the photon). The law of baryon conservation states that the total value of A, like electric charge, can never change in a reaction. Baryons cannot be

created or destroyed, except when a baryon-antibaryon pair annihilate each other or are created together.

The second conserved quantity is J, or spin angular momentum, which measures how fast a particle rotates about its axis. It is a fundamental feature of quantum theory that a particle can have a spin of only integral or half-integral multiples of Planck's constant. (This constant, ħ, relates the energy of a quantum of radiation to its wavelength: energy equals 2π times frequency times ħ.) For baryons J is always half-integral (that is, half an odd integer, such as 1/2, 3/2, 5/2 and so on) and for mesons J is always integral (that is, 0, 1, 2 and so on).

The third conserved quantity, closely associated with J, is P, or intrinsic parity. Parity is conserved when nature does not distinguish between left and right. Because such symmetry is observed in strong interactions, quantum mechanics tells us that an intrinsic parity value of + 1 or − 1 can be assigned to each strongly interacting particle. In the case of weak interactions, however, nature does distinguish between left and right, and the symmetry is violated.

The bookkeeping on parity is not quite so simple as that for electric charge and baryon number; the intrinsic parity values on each side of an equation are not necessarily the same. The reason is that the total parity is affected by spin angular momentum as well as by intrinsic parity. The close connection between parity and spin angular momentum makes it convenient to write the spin angular momentum quantum number J and the intrinsic parity P next to each other in describing each particle. For the proton, for example, J equals $1/2$ and P equals $+1$, which is shortened to J^P equals $1/2^+$. For the pion J equals 0 and P is -1, so that one writes J^P equals 0^-. (The system of bookkeeping for J is actually quite complicated in quantum mechanics, but the details need not concern us here.)

The fourth quantity conserved in strong interactions is I, or isotopic spin. This quantum number has nothing to do with spin or angular momentum, except that its peculiar quantum-mechanical bookkeeping is similar to that for J. The concept of isotopic spin was originally introduced into quantum mechanics to accommodate the fact that the nucleon exists in two charge states: one positively charged (the proton) and the other neutral (the neutron). As far as strong inter-

actions are concerned, these two states behave alike; they are related to each other by the symmetry of isotopic spin. Moreover, if the symmetry were observed by the electromagnetic interaction, the proton and the neutron would have the same mass. Precisely because isotopic-spin symmetry is violated by the electromagnetic interaction the neutron is 1.3 Mev (or .14 per cent) more massive than the proton.

A set of particles or particle states (we use the terms interchangeably) related to each other by isotopic-spin symmetry is a charge multiplet and is given a single name. Thus the nucleon doublet consists of the two charge states, positive and negative. The pion triplet consists of negative, neutral and positive charge states. The number of different charge states in a multiplet, or its "multiplicity" (M), is directly related to the isotopic-spin quantum number I by the equation $M = 2I + 1$. For the nucleon M equals 2 and I equals $1/2$; for the pion M equals 3 and I equals 1.

The fifth conserved quantity goes by any of three names: average charge ($\overline{Q}$), hypercharge (Y) or strangeness (S),

which are related to each other in a simple way. Average charge is just what its name implies: the average of the electric charges in a multiplet. Hence for the nucleon it is $1/2$ (0 plus 1 divided by 2); for the pion it is 0. Hypercharge is defined as twice the average charge (Y equals $2\overline{Q}$) merely to make it an integral number. And strangeness is hypercharge minus the baryon number (S equals Y minus A). It is clear that the three quantities are in effect interchangeable.

The concept of strangeness and its conservation is only 11 years old. In the early 1950's certain particles such as the K, the sigma and the xi were being observed for the first time, and because of their unusual behavior they were referred to as "strange particles." Most of them have relatively long lifetimes, which indicates that they decay by the weak interaction rather than by the electromagnetic or strong interaction. On the other hand, they are readily produced in high-energy collisions of "ordinary" particles (pions and nucleons), which proves that the strange particles too are strongly interacting. When invariable behavior patterns of this sort are observed, the

CONSERVED QUANTITY	SYMBOL	OBSERVED VALUES	DESCRIPTION	EXAMPLES	
				proton	negative pion
ELECTRIC CHARGE	Q	$0, \pm1, \pm2, \pm3 \ldots$	Represents the number of electric-charge units carried by a particle, or atomic nucleus, in units of the positive charge on the proton. Charge multiplets, such as the neutron-proton doublet or the pion triplet, can be assigned an average charge, $\overline{Q}$.	$Q = +1$ $\overline{Q} = +1/2$	$Q = -1$ $\overline{Q} = 0$
ATOMIC MASS NUMBER, OR BARYON NUMBER	A	$0, \pm1, \pm2, \pm3 \ldots$	Represents the familiar atomic mass number long used for nuclei. For uranium 235, $A = 235$. For baryons, $A = +1$; for antibaryons, $A = -1$; for mesons, $A = 0$.	$A = +1$	$A = 0$
HYPERCHARGE (Related to average charge, $\overline{Q}$, and to strangeness, S)	Y	$-2, -1, 0, +1$	Defined as twice the average charge, $\overline{Q}$, of a multiplet. Strangeness, S, is hypercharge minus the atomic mass number ($S = Y - A$).	$Y = +1$ $S = 0$	$Y = 0$ $S = 0$
ISOTOPIC SPIN (Related to multiplicity, M)	I	$0, 1/2, 1, 3/2$	Groups nuclear states into multiplets whose members differ only in electric charge. The number of charge states, or multiplicity, M, is related to I by the equation $M = 2I + 1$.	$I = 1/2$ $M = 2$	$I = 1$ $M = 3$
SPIN ANGULAR MOMENTUM	J	$1/2, 3/2, 5/2 \ldots$ $0, 1, 2, 3 \ldots$	Indicates how fast a particle rotates about its axis, expressed in units of Planck's constant, $\hbar$.	$J = 1/2$	$J = 0$
PARITY	P	$-1, +1$	An intrinsic property related to left-right symmetry.	$P = +1$	$P = -1$
G	G	$-1, +1$	An intrinsic property found only in mesons with zero hypercharge.	not defined	$G = -1$

CHART OF QUANTUM NUMBERS shows seven quantities conserved by the strong interaction but not necessarily by the electromagnetic or weak interaction. The three quantities in color (A, Y, I) are easily established by experiment and provide the basis for assigning a family name to each particle. Only 10 combinations of A, Y and I are now known, each represented by a Greek letter.

physicist suspects that a conservation law (or symmetry) is at work. One of the authors of this article (Gell-Mann) and the Japanese physicist Kazuhiko Nishijima independently proposed that a previously unsuspected quantity (strangeness, or hypercharge) is conserved in strong and electromagnetic interactions but violated by weak interactions. The hypothesis made it possible to predict the existence and general properties of several strange particles before they were discovered.

The New Nomenclature

We are now ready to describe how the five quantum numbers can provide the basis for a new naming system. By appropriate selection of three of the five quantum numbers we can indicate immediately whether a strongly interacting particle is a baryon or a meson, how many members it has in its immediate family (that is, its multiplicity) and what its degree of strangeness is. The three quantum numbers that provide this information are the atomic mass, or baryon, number A, the hypercharge Y and the isotopic spin I. (It will be recalled that Y is directly related to strangeness and I to multiplicity.)

Now, partly as a mnemonic aid and partly out of respect for the old pet names, we shall employ a letter symbol to indicate various combinations of A, Y and I. To designate the known mesons, particles for which A is 0, it is sufficient to use four lower-case Greek letters: η (eta), π (pi), κ (kappa) and $\bar{\kappa}$ (antikappa, or kappa bar). The chart at the bottom of the next page shows the Y and I values for each symbol. Even though the multiplicity M can be found simply by doubling I and adding 1, it is shown separately for easy reference.

To designate baryons, for which A is 1, we will use the following upper-case Greek letters: Λ (lambda), Σ (sigma), N (which stands for the nucleon and is pronounced "en," not "nu"), Ξ (xi), Ω (omega) and Δ (delta). The values of Y, I and M for each symbol are also shown at the bottom of the next page.

These 10 symbols encompass all the kinds of meson and baryon states that are known at the present time. In other words, any of the 82 particles in the chart on pages 40 and 41 can be designated by one of these 10 symbols. The difference between the old naming system and the new one should now be apparent. In the old system, shown in the chart on page 39, the symbol π, for example, represented only a single family of three particles with a rest mass of 137 Mev. In the new system π represents both that group and a new group of three, with identical A, Y and I but with a rest mass of 750 Mev. Similarly, in the old system N stood only for the nucleon doublet with a mass of 939 Mev. In the new system N stands for the nucleon and for two higher energy states, also doublets, one with a mass of 1,512 Mev and one with a mass of 1,688 Mev. Thus the old particle names now stand for classes of particles with the same A, Y and I.

Various members of a class can be

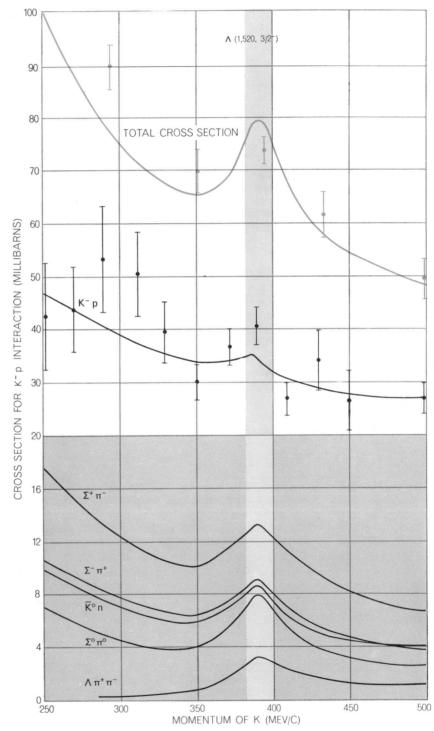

MULTIPLE DECAY MODES of $\Lambda(1,520, 3/2^-)$ show that the baryon Λ is connected to six open channels. The Λ is created when K$^-$ mesons [$\kappa(496)$] are scattered on protons. The total interaction cross section shows a peak corresponding to the formation of $\Lambda(1,520, 3/2^-)$, which speedily decays into one of the six channels indicated. Thus the curve labeled K$^-$p represents the over-all reaction: K$^-$p $\rightarrow \Lambda \rightarrow$ K$^-$p. Thousands of bubble-chamber "events" supplied the data for the curves. Data points, however, are given only for the top two.

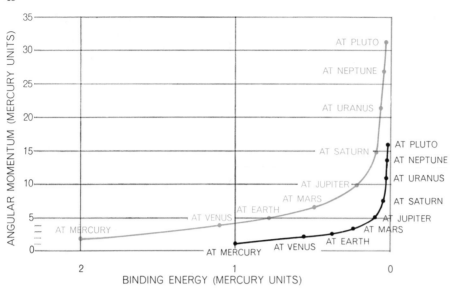

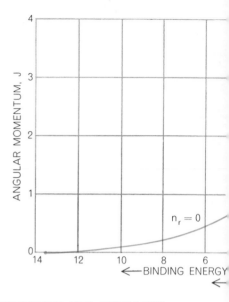

"REGGE TRAJECTORIES," an important concept for predicting recurrences, can be explained with a spaceship analogy. If one-ton spaceships were placed in circular orbits around the sun at the distance of each of the nine planets, the ships would have the binding energies and angular momenta indicated. The black curve drawn through these values is a Regge trajectory. The colored curve is a Regge trajectory for a two-ton spaceship.

TRAJECTORIES FOR HYDROGEN resemble those for a spaceship. The hydrogen atom, made up of an electron "circling" a proton, can exist in various states of excitation. For each value of a certain quantum

---

distinguished by writing the mass value in parentheses after the name, for example $\pi(137)$ or $\pi(750)$; or by writing the angular momentum J and the parity P in parentheses, for example $\pi(0^-)$ or $\pi(1^-)$. Or, if desired, mass, J and P can all be shown: $\pi(137, 0^-)$ or $\pi(750, 1^-)$. In the following discussion an unadorned symbol will refer to the lowest mass state in the class. The new classification system is exhibited in the baryon and meson charts on page 42.

Particle Stability

We mentioned earlier that particles can decay in one of three ways: via the strong, the electromagnetic or the weak interaction. A few particles (the photon, the two neutrinos, the electron and the proton) are absolutely stable provided that they do not come into contact with their antiparticles, whereupon they are annihilated. Particles that decay through the electromagnetic or weak interaction are said to be metastable. Those that decay through the strong interaction are called unstable and have very short lifetimes, typically a few times 10^{-23} second. This is still a considerable length of time, however, compared with less than 10^{-23} second, which is the characteristic time required for a collision between high-speed particles.

Unstable particles are those with enough energy to decay into two or more strongly interacting particles without violating any of the conservation laws respected by the strong interaction. Some unstable particles have only one possible decay mode; others have more than one. As an example of the former, $\Xi(1,530)$ decays only into Ξ and π. On the other hand, $\Lambda(1,520)$ can decay into Σ and π, into N and $\bar{\kappa}$ or into Λ and two π's.

How can the existence of several decay modes be explained? This can be answered by introducing the concept of "communicating states." A nuclear state can be either a single particle or a combination of two or more particles. We have seen that each particle has definite values of the conserved quantum numbers A, Y, I, J, P and, where applicable, G. The strong interaction allows transitions, or communication, only among nuclear states with the same values for all the conserved quantum numbers.

Now, one can write down many nuclear states, consisting of two or more particles, that have in the aggregate the same set of quantum numbers as any particular unstable particle. For decay to take place, however, the unstable particle must have a rest mass at least equal to the threshold energy (that is, the sum of the rest masses) of the particles into which it might conceivably decay. In other words, energy must be conserved. The various states into which an unstable particle has sufficient energy to decay are called open channels. Communicating states with threshold energies greater than that available to the unstable particle are called closed channels; decay into them is allowed by everything *but* conservation of energy. A schematic representation of some of the channels that communicate with $\pi(750, 1^-)$ is shown on page 43.

This leads us to the concept of "resonance," the term originally applied to unstable particles. The first of these was discovered in 1952 at the University of Chicago by Enrico Fermi and his colleagues. At the time no one suspected that a deluge was to follow.

MESONS	Y	I	M
η	0	0	1
π	0	1	3
κ	+1	1/2	2
κ̄	−1	1/2	2

BARYONS	Y	I	M
Λ	0	0	1
Σ	0	1	3
N	+1	1/2	2
Ξ	−1	1/2	2
Ω	−2	0	1
Δ	+1	3/2	4

SYMBOLS FOR MESONS AND BARYONS are Greek letters. For mesons atomic mass number A is 0; for baryons it is 1. The 10 letters identify the 10 known combinations of mass number (A), hypercharge (Y) and isotopic spin (I). Multiplicity (M) is related to I.

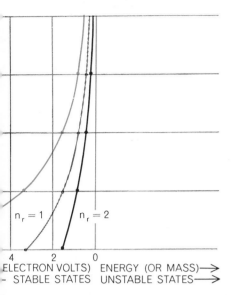

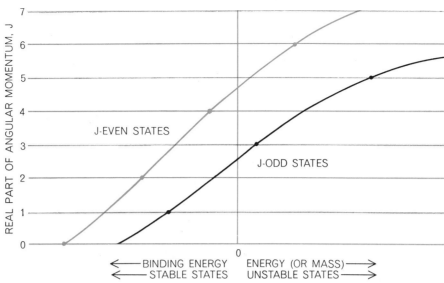

REAL PART OF ANGULAR MOMENTUM, J

$n_r = 1$ $n_r = 2$

4   2   0
ELECTRON VOLTS)   ENERGY (OR MASS)→
– STABLE STATES   UNSTABLE STATES→

J-EVEN STATES

J-ODD STATES

←—BINDING ENERGY   ENERGY (OR MASS)—→
←——— STABLE STATES   UNSTABLE STATES——→

number (n_r) the binding energies of the hydrogen states decrease with increasing values of angular momentum (J). Smooth curves drawn through these states (each a different "particle") are Regge trajectories.

TRAJECTORIES FOR STRONGLY INTERACTING PARTICLES resemble those at left except that they include unstable states. An intersection with J in the stable region indicates the existence of a stable or metastable particle. Intersections in the unstable region indicate an unstable particle. Trajectories connect particle states separated by two full units of J. Lowest state is an "occurrence"; higher states are "Regge recurrences."

In the 1952 experiments pions from the University of Chicago cyclotron were directed at protons (in liquid hydrogen) and the scattering cross section was measured for different energies of the pion beam. Scattering refers to the change of direction when two particles collide; the scattering cross section is the probability that scattering will occur. When the probability is large, the two particles act as if they were big, with a large cross section.

For each setting of the pion beam the "effective" mass of the pion-proton system is calculated. The effective mass is the sum of the rest masses and the kinetic energies of all the particles in a system, as viewed from the system's center of mass. When the effective mass is plotted along one co-ordinate of a graph and the scattering cross section along the other, it is found that the cross section peaks sharply when the system has an effective mass of about 1,238 Mev. It is through this peak that the resonance was detected. We shall discuss below the connection between the peak and the resonance, or unstable particle, that is called $\Delta(1,238)$ in our new notation.

The University of Chicago cyclotron could not create pion-proton systems with an effective mass of much more than 1,300 Mev. Subsequently, with more powerful accelerators, it was found that pion-proton scattering produces a whole series of resonances. The illustration on page 45 shows two resonances created when positive pions are

scattered by protons and, in a separate curve, four resonances created when negative pions are scattered by protons. The first two resonances are $\Delta(1,238, 3/2^+)$ and $\Delta(1,920, 7/2)$. The same two should also show up in negative pion scattering, but in the actual experimental curve for the negative pion the upper one, $\Delta(1,920, 7/2)$, is hardly observable. We do see, however, two other resonances, $N(1,512, 3/2)$ and $N(1,688, 5/2)$, that cannot contribute to the scattering of positive pions by protons. (A space in the parentheses following the half-integer value of J indicates that the parity of the particle has not yet been established.)

Although $\Delta(1,238)$ decays mainly into one pion and a nucleon, the higher resonances can also decay into two or more pions and a nucleon. Most resonances can decay in more than one way; that is, they can communicate with several open channels [see illustration on page 47].

To explain how an unstable particle can communicate with several open channels we have found it helpful to draw an analogy between the behavior of unstable particles and the behavior of resonant cavities such as organ pipes and electromagnetic cavities. Cavities of the latter sort (such as the magnetron tube employed in radar) are used in electronics to create intense electromagnetic waves of a desired frequency, which is a resonant frequency of the cavity. Each cavity has a characteristic "lifetime": the time required for the

electromagnetic radiation to leak out.

In quantum mechanics, particles and waves are complementary concepts, and the amount of energy associated with a particle, or nuclear state, can be expressed as an equivalent frequency. In other words, energy is proportional to frequency. The fact that the Δ particle appears when a pion is scattered by a proton at or near a certain energy —the resonance energy—is equivalent to saying that the particle appears at a certain frequency. Thus a resonance energy in particle physics can be compared to the resonance frequency of an acoustic or electromagnetic cavity. What is the "cavity" in particle physics? It is an imaginary structure: one cavity, each with its own special properties, for each set of values of the quantum numbers conserved in strong interactions.

The analogy between unstable particles and the resonant modes of electromagnetic cavities can be carried further. To the electromagnetic cavity one can attach the long pipes known as wave guides, which have the property of efficiently transmitting electromagnetic waves of high frequency but not those of low frequency. When the electromagnetic wavelength is slightly larger than the dimensions of the wave guide, the guide refuses to transmit. In this sense the wave guide acts like a particle channel that is open only above its characteristic threshold energy. If a cavity has attached to it several wave guides of different sizes, high-frequency radiation can flow into the cavity through one guide and flow out through the same or different guides.

By analogy energy can flow into a nuclear interaction through one channel and pass out through one or more open channels. As the energy (frequency) is increased from low values, the channels open up one by one and new nuclear reactions become possible, with energy going out through any of the open channels. Now, as the frequency is increased, suppose it passes through a resonance frequency of the nuclear cavity. At this point it becomes easier for the cavity to absorb and reradiate energy. The resonance appears as a peak in the scattering cross section of a nuclear reaction. In other words, a resonant mode of the cavity corresponds to an unstable particle, such as Δ or $\pi(750)$.

Just as an electromagnetic cavity that is near resonance holds on to electromagnetic energy for a long time, so the unstable particle typically takes somewhat more than the characteristic time of less than 10^{-23} second to decay. If energy is fed into the cavity through one pipe, stays in the cavity for a while because of resonance and comes out again through the original pipe, that corresponds to a scattering collision between two $\pi(137)$ particles that produce the unstable particle $\pi(750)$, which finally decays again into the original particles.

Alternatively, the energy can emerge through another pipe, which corresponds to the case in which $\pi(750)$ decays into four $\pi(137)$ particles. These, of course, are only two of many examples. The cavity and wave guide analogies are illustrated on page 44.

One can use the wave guide analogy to describe not only unstable particles but also stable ones. A stable particle is merely one that has such a low mass that all the communicating channels are closed. Therefore it is a "bound" state rather than a scattering resonance. For an electromagnetic cavity this condition would correspond to a resonant mode whose frequency is below the threshold frequency of all the wave guide outlets. If radiation could be put into the cavity in such a mode, it could not leak out. Of course, an actual cavity would eventually lose radiation by leakage into and through its walls. Such leakage corresponds to the decay of metastable particles via the weak and electromagnetic reactions. An absolutely stable particle really does live forever.

The reader who is unfamiliar with the phenomenon of resonance in electromagnetic cavities may be wondering if we have simplified his task by introducing the electromagnetic analogy. Would it not be just as easy to explain resonances in particle physics directly? Possibly so. But by drawing attention to

similar behavior in two apparently different fields we hope we have illustrated a unity in physics that may make particle behavior seem less esoteric. The more basic value of the analogy, however, is that it has helped theorists to understand some deeper points in particle resonances than we have been able to talk about here.

Regge Trajectories

As the strongly interacting particles proliferated, physicists sought to find patterns that would show relations among them. In particular they tried to find classification systems that might predict new particles on the basis of those already known. The first concept to prove useful in this respect developed from an idea introduced into particle physics in 1959 by the Italian physicist Tullio Regge. The concept already had counterparts throughout quantum physics, notably in the study of atomic- and nuclear-energy levels.

It had been observed that as particles increase in mass they frequently (but not invariably) exhibit a higher value of spin angular momentum J. Regge pointed out the existence, in many important cases, of a mathematical relation between the value of J and particle mass. He showed that certain properties of particles can be regarded

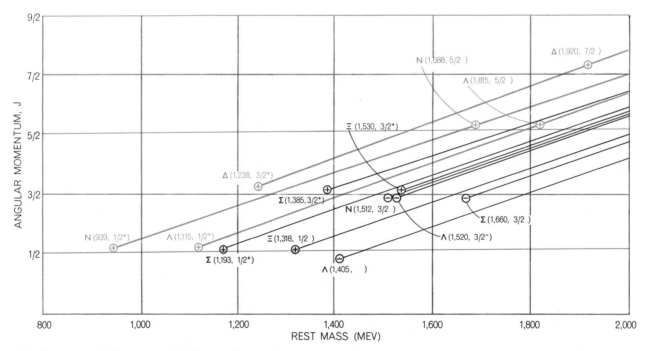

REGGE TRAJECTORIES FOR BARYONS are shown for 14 well-established baryons with mass less than 2,000 Mev. Slanting colored lines connect three occurrences with their Regge recurrences. For baryons spin angular momentum (J) is half-integral (1/2, 3/2, 5/2 and so on). Recurrences must have 2, 4, 6 and so on more units of spin than their ground states (occurrences) of lowest mass. Spins for $\Lambda(1,815, 5/2)$ and $\Delta(1,920, 7/2)$ are uncertain, but they probably satisfy this requirement. Slanting black lines show the probable Regge trajectories for other baryons. Circled symbols indicate parity; where not yet established, it has been guessed.

as "smooth" mathematical functions of J, that is, functions that vary continuously as J varies. But since in quantum mechanics J can have only integral and half-integral values, the functions have direct physical meaning only for those permitted values. The smooth mathematical curve that gives the physical mass for different values of J is called a "Regge trajectory."

A spaceship analogy may help to clarify the concept of the Regge trajectory. Suppose in the nearly circular orbit occupied by each of the sun's nine planets one were to place a one-ton spacecraft. These craft circle the sun as if they were miniature planets. The nearer a craft is to the sun, the more strongly it "feels" the sun's gravitational force and the more strongly it is bound. This binding energy, therefore, is highest for the craft in Mercury's orbit and least for that in Pluto's orbit. (The binding energy is just that amount needed to release the craft from the sun's attractive force.)

Each craft can be assigned another quantity that also varies with its distance from the sun: angular momentum. It frequently happens in physics that, other things being equal, the greater the binding energy, the smaller the angular momentum. In our example this means that angular momentum increases with the distance from the sun.

One can now draw a graph for the nine spacecraft in which angular momentum is plotted on the vertical axis and binding energy on the horizontal axis [see *illustration at top left on page 48*]. The curve drawn through the plotted points is analogous to a Regge trajectory.

Now, suppose quantum mechanics were to have a controlling effect on the macroscopic scale of spacecraft and solar orbits. Suppose, that is, the angular momentum of the spacecraft in Mercury's orbit represented an elementary quantum of spin. If such were the case, a one-ton craft would be allowed to occupy only those orbits in which the angular momenta (expressed in Mercury units) assumed integral values. This is equivalent to saying that a one-ton craft in a circular orbit could exist only at certain energy levels. The Regge trajectory for the spaceship would then be physical only at these points. Another Regge trajectory could be drawn for a two-ton spacecraft. In any given circular orbit its binding energy and angular momentum would be twice that of a one-ton spacecraft.

Although it was not used by physicists until recently, the notion of a Regge trajectory applies to problems long familiar in atomic physics. It is well known, for example, that the electron-proton system constituting the hydrogen atom can exist in various states of ex-

citation. The electron can occupy various orbits around the proton, just as the spaceship could occupy many orbits around the sun. In the case of the electron, of course, the quantization of the orbits is very conspicuous. When the value of a certain quantum number, n_r (a number characterizing the energy of motion in a radial direction), is held fixed, the binding energies of the various hydrogen states decrease with increasing values of the angular momentum J. If a smooth curve is drawn through the permitted values of J, one obtains a Regge trajectory similar to that connecting the spacecraft in different orbits [see *middle illustration at top of pages 48 and 49*]. For each value of n_r in the hydrogen atom there is a different trajectory, just as there is for each spacecraft of different mass.

In the case of the hydrogen atom the intersection of a Regge trajectory with a permitted value of J (0, 1, 2 and so on) corresponds to the occurrence of a bound state. From an experimental standpoint each of these occurrences is a different "particle" with a different mass. The series of occurrences is brought to an end when the excitation energy becomes so great that the electron is dissociated from the proton. This energy limit divides stable states from unstable states.

Just as Regge trajectories can be

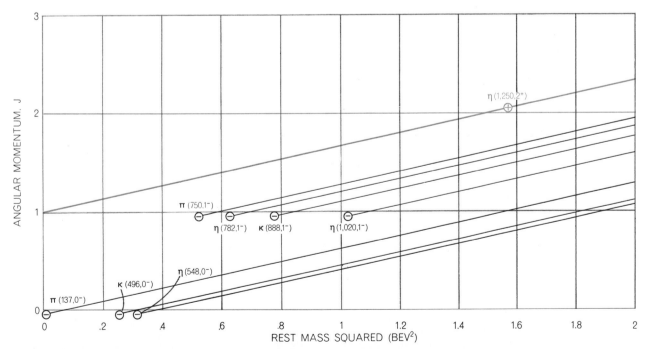

REGGE TRAJECTORIES FOR MESONS are shown for eight well-established particles. They are all in the ground state; no recurrences have yet been identified. It can be shown that the highest η trajectory (that is, a trajectory with a Y and I of 0) should have an unreal intersection at J of 1 and mass 0. A colored line drawn through that point and $\eta(1,250, 2^+)$ indicates the probable slope of Regge trajectories for mesons. The parallel black lines for other trajectories are hypothetical. Their intersections with a J of 2 or 3 predict where Regge recurrences are likely to be found. The lowest-lying recurrence should be a $\pi(2^-)$ of about 1,700 Mev.

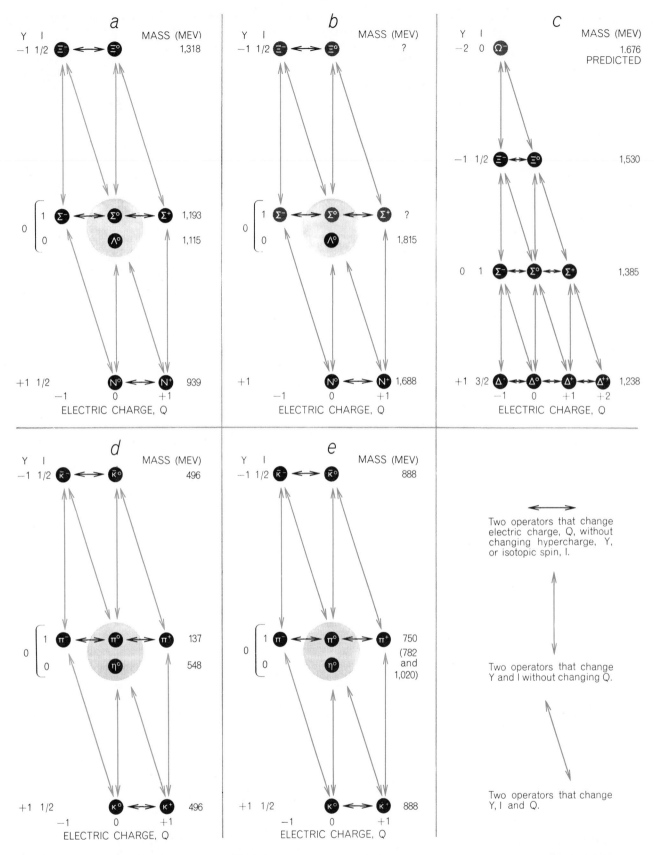

"EIGHTFOLD WAY" invokes a new system of symmetries to group multiplets of particles into "supermultiplets." The term "eightfold" refers to a special algebra showing relations among eight things, in this case eight conserved quantities. The new system of symmetries (*colored arrows*) connects different values of hypercharge (Y) and isotopic spin (I) in the same way that isotopic-spin symmetry (*black arrows*) connects different values of electric charge. Four of the diagrams (*a, b, d, e*) show supermultiplets with eight members; another group (*c*) contains 10 members. Several new particles are predicted by the eightfold way, notably $\Omega(1,676, 3/2^+)$, which appears in *c*. Note that the η meson in *e* is given two mass values, which leads to the "identity crisis" described in the text.

drawn for the gravitational force (spaceship example) and the electromagnetic force (hydrogen-atom example), trajectories can be drawn for the strong force governing strongly interacting particles. In this case the trajectories do not terminate at the boundary between stable and unstable states but continue across, cutting further integer values of J [see *illustration at top right on page 49*]. An intersection in the stable region indicates the existence of a bound state, meaning a particle that is either stable or metastable. Intersections in the unstable region indicate the existence of resonances, or unstable particles. It can be shown that for strongly interacting particles a particular trajectory joins up real states for either odd or even values of J but not for both. This means that an interval of two units of J must intervene between states on the same trajectory. The lowest state is called an occurrence; higher states can be referred to as Regge recurrences, or as a series of excited rotational states.

How can the existence of a trajectory be demonstrated? In analogy with the spaceship or hydrogen-atom example, one plots the angular momentum J against mass (in Mev) for all the particles that share all the same quantum numbers except J. One can then quickly see whether or not they fall into groups that lie on rising curves. If they do, one has an indication of Regge trajectories. Such trajectories for baryons are illustrated on page 50.

The rule says that only states separated by two full integers can lie on the same trajectory. So far three pairs of states appear to meet this requirement: the two N states, N(939, $1/2^+$) and N(1,688, $5/2$); the two Λ states, Λ(1,115, $1/2^+$) and Λ(1,815, $5/2$), and the two Δ states, Δ(1,238, $3/2^+$) and Δ(1,920, $7/2$). (The spins of the higher Λ and Δ states are not certain; they may be greater than $5/2$ and $7/2$ respectively.)

In the illustration on page 50 these three pairs of states are connected by slanting colored lines. The black lines represent assumed trajectories on which only one occurrence has so far been definitely discovered. These conjectured trajectories are useful in telling experimenters where to search for baryons of higher angular momentum.

The adjacent illustration on page 51 shows Regge trajectories plotted for mesons. For mesons quantum mechanics tells us to take mass squared rather than mass for the horizontal scale. It will be seen that no Regge recurrences have

been discovered as yet, perhaps because meson states of high mass have not yet been studied carefully.

The best evidence for the existence of a Regge trajectory among mesons is based on certain arguments showing that a particular Regge trajectory for a meson state with a Y of 0 and an I of 0 should have an unphysical, or unreal, intersection with a J of 1 and a rest mass of 0. The next lower intersection, at a J of 0, could be physical except that the state would have negative mass squared, which has no meaning. Thus the lowest real intersection should occur at two units of J above 0, that is, at a J of 2. In fact, a meson with a J of 2, designated $\eta(1,250, 2^+)$, has apparently been discovered within the past year and a half. Its quantum numbers are still uncertain, however. If a Regge trajectory is drawn between $\eta(1,250, 2^+)$ and its unphysical intersection at a J of 1 and a rest mass of 0, one obtains a crude indication of the slope for the other meson trajectories, shown by the black lines in the illustration. Vigorous experimental efforts are under way to find second members of these meson families, with a J of 2 or 3.

The Eightfold Way

Now we shall turn to another classification scheme that has proved valuable in predicting the existence of previously undiscovered particles. We have seen how the notion of Regge trajectories makes it possible to perceive family connections between particles with different values of J but the same values of all other quantum numbers. Now we shall describe a relation that seems to exist among particles with the same values of J and of parity P but different values of mass, hypercharge Y and isotopic spin I.

We mentioned earlier that the difference in mass between charge multiplets such as the nucleon doublet (neutron and proton) can be regarded as a "splitting" caused by the fact that isotopic spin is not conserved by the electromagnetic interaction, which underlies the electric charge. This violation produces a maximum mass difference of about 12 Mev in the case of the Σ triplet.

Now, it is a remarkable fact that the four best-known members of the baryon family, N, Λ, Σ and Ξ, are separated by average mass differences only about a factor of 10 greater than that separating members within each multiplet. The gaps in average mass separating the four baryon states are only 77, 75 and 130 Mev respectively. Moreover, these four baryons all seem to have the same J^P; it

is $1/2^+$. (Actually the J of Ξ is not firmly established and its parity is still unmeasured.)

If the difference in mass within a multiplet is caused by a violation of the isotopic spin I, is it conceivable that the somewhat greater difference in mass between neighboring multiplets is caused by the violation of the conservation of some other quantum numbers? The kind of solution needed is one in which Y and I are exactly conserved by the strong interaction but certain other conservation laws are broken by some aspect, or some part, of this same interaction. If such a partial violation of new symmetry principles were permitted, one might be able to group baryon multiplets into "supermultiplets" with various values of Y and I but the same J and P. This new system of symmetries would connect different Y and I values in the same way that isotopic spin connects different values of electric charge. The aspect of the strong interaction that violates the new symmetries—represented by new quantum numbers—would split each supermultiplet into charge multiplets of different mass, much as the electromagnetic interaction causes splitting of mass among the members of a charge multiplet by violating isotopic-spin symmetry. The scale of the mass splitting within the supermultiplet, however, would be much greater than that observed within a multiplet, since it is an appreciable fraction of the strong force that is at work rather than the electromagnetic force, which is much weaker.

Early in 1961 an Israeli army colonel and engineer-turned-physicist, Y. Ne'eman, and one of the authors (Gell-Mann), working independently, suggested a particular unified system of symmetries and a particular pattern of violations that made plausible the existence of supermultiplets. The new system of symmetries has been referred to as the "eightfold way" because it involves the operation of eight quantum numbers and also because it recalls an aphorism attributed to Buddha: "Now this, O monks, is noble truth that leads to the cessation of pain: this is the noble *Eightfold Way:* namely, right views, right intention, right speech, right action, right living, right effort, right mindfulness, right concentration."

The mathematical basis of the eightfold way is to be found in what are called Lie groups and Lie algebras, which are algebraic systems developed in the 19th century by the Norwegian mathematician Sophus Lie. The simplest Lie algebra involves the relation of three components, each of which is a symmetry operation of the kind used in quantum mechanics. Isotopic spin consists of three such components (I_+, I_- and I_z) related by the rules of this simplest algebra. The algebra is that of the Lie group called SU(2), which stands for special unitary group for arrays of size 2×2; there is one condition in the 2×2 arrays that reduces the number of independent components from four to three (hence the term "special").

The component operations of the eightfold way satisfy the mathematical relations of the next higher Lie algebra, which has eight independent components. Here the Lie group is called SU(3), which stands for special unitary group for arrays of size 3×3; again a special condition reduces the number of components from nine to eight. The eight conserved quantities of the eightfold way consist of the three components of isotopic spin, the hypercharge Y and four new symmetries not yet formally named. Two of the new symmetries change Y up or down by one unit without changing electric charge; the other two symmetries change both Y and electric charge by one unit [see illustration on page 52]. The violation of all four new symmetries by part of the strong interaction changes the masses of the multiplets forming a supermultiplet. An example of a supermultiplet (an octet) is provided by N, Λ, Σ and Ξ, if indeed they all have an angular momentum J of 1/2 and positive parity, that is, a J^P of $1/2^+$.

The kind of violation suggested by the eightfold way leads to a rule connecting the masses of a supermultiplet, provided that the violation is not too severe. The rule for N, Λ, Σ and Ξ is that 1/2 mass N plus 1/2 mass Ξ equals 3/4 mass Λ plus 1/4 mass Σ. Substituting the actual masses of the four particle states gives 1,129 Mev for the left side of the equation and 1,135 Mev for the right side. The agreement with the approximate mass rule is surprisingly good.

This apparent success suggested a search for other octets. At the beginning of 1961 the only meson multiplets certainly known were π, κ and $\bar{\kappa}$, all with a J^P of 0^-. They just fit the octet pattern if one adds a neutral singlet meson with a predicted mass of 563 Mev. (The masses are predicted as they are for baryons except that for mesons the masses in the equation must be squared.) Late in 1961 the η meson was found, with a mass of 548 Mev. Later its J^P was determined to be 0^-, as required.

Meanwhile, mesons with a J^P of 1^- were turning up: the $\pi(750, 1^-)$ triplet

and the $\kappa(888, 1^-)$ and $\bar{\kappa}(888, 1^-)$ doublets. Once more the octet pattern was appearing, and the existence of a neutral singlet with a mass of about 925 Mev was expected. Very soon experiments revealed the $\eta(1^-)$ meson, but its mass was only 782 Mev. The mass rule, so successful before, had mysteriously failed.

The mystery may since have been cleared up somewhat. The octet is only one of several supermultiplets allowed by the eightfold way. Another possibility is a lone neutral singlet with a Y of 0 and an I of 0. Suppose there were such a meson with a J^P of 1^-, which we call $\eta'(1^-)$. If its mass value were near that of $\eta(1^-)$, only the broken symmetries of the eightfold way would distinguish the two particles. Under these conditions quantum mechanics predicts that a kind of "identity crisis" would set in, making each of the mesons assume to some extent the properties of the other one. Moreover, the masses would be affected by this sharing of properties, and one meson would have a higher mass, the other a lower mass, compared with the simple case in which the rule for the octet holds. The predicted mass squared for $\eta(1^-)$, or $(925)^2$, should then lie roughly halfway between the actual values of mass squared for $\eta(1^-)$ and $\eta'(1^-)$. Since the actual mass of $\eta(1^-)$ is 782 Mev, one might expect to find another meson, $\eta'(1^-)$, with a mass of around 1,045 Mev. Indeed, in 1962 such a meson with the right values of Y and I (both 0), and with a mass of 1,020 Mev, was discovered independently by two groups of physicists. There is actually no clear way to decide which of the two η mesons, $\eta(782)$ or $\eta(1,020)$, belongs to an octet and which is the singlet. We assume that nature is as perplexed as we are.

The $\eta(1,250, 2^+)$ meson, the most recently found of the mesons shown in the chart on pages 40 and 41, seems to be a singlet. Thus the 18 mesons listed can be accounted for as two octets and two singlets. Not all the experimental facts are certain, however, and the picture, particularly the identity crisis of $\eta(1^-)$ and $\eta'(1^-)$, must still be considered tentative.

Returning to the baryons, what other supermultiplets have been found beyond the original one containing N, Λ, Σ and Ξ? $\Lambda(1,405)$ seems to be a singlet; its J^P has not been definitely established. N(1,688, 5/2), the first Regge recurrence of the nucleon, should, like the nucleon, belong to an octet containing other Regge recurrences of Λ, Σ and Ξ. The Λ member of this excited octet may well be $\Lambda(1,815)$, if indeed it has a J^P

of $5/2^+$. The Σ and Ξ members are now being sought; if one of the two particles can be found, the mass of the other can be predicted approximately by the octet mass rule.

N(1,512, 3/2) may also belong to an octet. Another probable member has already been found: $\Lambda(1,520, 3/2^-)$. It is possible that $\Sigma(1,660)$ has a J^P of $3/2^-$. If this assignment is correct, the octet mass rule predicts a Ξ multiplet with a mass around 1,600 Mev. Here, however, the experimental situation is very uncertain.

That brings us to $\Delta(1,238)$, the unstable baryon discovered in 1952. Since it is a quartet, it cannot belong to either the octet or the singlet pattern. The simplest supermultiplet permitted by the eightfold way into which it can fit is a 10-member group, or decuplet, consisting of a Δ quartet, a Σ triplet, a Ξ doublet and an Ω singlet [see "c" in illustration on page 52]. For decuplets the mass rule predicts approximately equal mass spacing between members of the supermultiplet. Since $\Sigma(1,385)$ is thought to have a J^P of $3/2^+$, it could very well belong to a decuplet with $\Delta(1,238)$. The equal-spacing rule predicts a Ξ particle at about 1,532 Mev. The discovery of $\Xi(1,530)$, with a J^P probably of $3/2^+$, appears to be a striking confirmation of the prediction. The mass rule further predicts an Ω particle at about 1,676 Mev, which would be the only particle state consisting of a negative charge singlet. Such a particle would actually be stable under strong and electromagnetic interactions, since it would lack the energy to decay into any of the channels with which it communicates. Thus it should live about 10^{-10} second and decay by weak interactions. It is now being sought eagerly. If it is found, the correctness of the eightfold way will be strikingly established.

We close this section with the remark that the symmetry game may not yet be finished for strongly interacting particles. For example, there might exist some undiscovered quantum number that is conserved by the strong interactions and that has the value 0 for all known particles. Before strange particles were discovered, the strangeness quantum number (equivalent to Y) was of this kind. Experiments at very high energies with the next generation of accelerators might produce a similar situation with respect to an entirely new quantum number.

Composite Particles

The meaning of the term "elementary particle" has varied enormously as man's view of the physical universe has become more detailed. In the past few years it has become increasingly awkward to consider several scores of particles as elementary. Evidently a reappraisal of the entire elementary-particle concept is in order.

Let us begin by asking why we feel sure that certain particles such as the hydrogen atom are *not* elementary. The answer is that even though these particles have properties qualitatively similar to those of neutrons, protons and electrons, it has been possible theoretically to explain their properties by assuming that they are composites of other particles.

The hydrogen atom itself provides an outstanding example of what is meant by composite, because its properties have been theoretically predicted with enormous accuracy. It is important to realize that the hydrogen atom is not exactly composed of one proton and one electron. It is more accurate to say that it is so composed *most* of the time. The ground state of the hydrogen atom is a stable "particle" that communicates (via strong, electromagnetic and weak interactions) with a great variety of closed channels, of which electron plus proton is the most important. According to quantum mechanics any state consists part of the time of each of the channels that communicate with it. As an example, for a certain small fraction of the time the ground state of the hydrogen atom consists of the expected electron and proton plus an electron-positron pair. The effect of this channel on the energy of the atom is tiny, but it has been calculated and measured; the agreement is excellent. There are infinitely many other closed channels that contribute to the structure of the hydrogen atom, but fortunately their effect is negligibly small.

In strongly interacting systems complicated channels are more important. For example, the properties of the deuteron (A equals 2) have been predicted, assuming that this particle is a composite of neutron and proton, but here the accuracy of the predictions is much poorer than it is for the hydrogen atom because the effect of additional channels (involving pions, say) is substantial. Nevertheless, there is a general belief that, since the simplest channel accounts for the bulk of the observed properties of the deuteron, it should eventually be possible to improve the predictions systematically by inclusion of more channels. The same kind of statement can be made for all nuclei heavier than the deuteron, and there is no disposition to re-

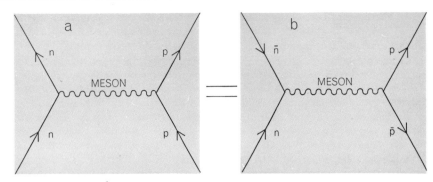

"CROSSED REACTIONS" illustrate the close correspondence between the concept of a force and the concept of a particle. Reaction *a*, which is read upward, represents a scattering collision between a neutron and a proton. The meson "exchanged" represents the strong force acting between two baryons. Reaction *b*, which is read from left to right, is a crossed reaction of *a*. It shows a meson acting as an intermediate particle in a reaction that converts a neutron and antineutron into a proton and antiproton. The two reactions are equivalent.

gard any of these compound nuclei as "elementary."

Confusion about the distinction between composite and elementary particles has arisen for particles with an A of 0 and 1 (mesons and baryons) because here one rarely has a single dominant channel nearby in energy. Consider one of the worst cases: the pion. The communicating channel with the lowest threshold is a 3π configuration; some channels with still higher thresholds are 5π, κ plus $\bar{\kappa}$ plus π, N plus $\bar{\text{N}}$, and Ξ plus $\bar{\Xi}$. Hence part of the time the pion exists as 3π, part of the time as κ plus $\bar{\kappa}$ plus π, and so forth.

All the relevant thresholds are much higher than the pion mass, and many rather complicated closed channels contribute substantially to the pion state. The result is that even a rough calculation of pion properties has not yet been achieved. A more favorable case is that of $\pi(750)$, where the 2π channel is believed to dominate; but even here a glance at the illustration on page 7 shows that there are many nearby channels to be reckoned with.

We may nevertheless employ the operational definition: a particle is nonelementary if all its properties can be calculated *in principle* by treating it as a composite. Such a calculation must yield various probabilities for the various closed channels; the binding forces in these channels must yield the right mass for the particle.

The problem of including all the significant channels is in most cases still too difficult, but suppose the calculation could be carried out. Would we then get a correct description of each particle? Would the quantum numbers and the mass come out right? Until recently there was an almost universal belief that a few strongly interacting particles, including

the nucleon, could not be calculated on such a basis. In the present theory of electrons and photons, which gives such an excellent description of electromagnetic phenomena, the properties of the electron and photon cannot be dynamically predicted. The reason is that the known forces are not powerful enough to form bound states with masses as small as those of the electron and the photon.

Reasoning by analogy, theorists tended to give the nucleon a special status parallel to that of the electron. Thus they were inhibited from trying to treat the nucleon as a composite particle. Gradually, however, this select status seemed increasingly dubious. And when an attempt was finally made to calculate the properties of the nucleon from an analysis of its communicating channels, the same qualitative success was achieved as with the deuteron and the $\Delta(1,238)$ particle, which had for years been called composite just because it had first been observed in a pion-proton scattering experiment.

It seems, furthermore, on the basis of recent developments in which the concept of the Regge trajectory plays an important role, that in all such dynamical calculations no distinction need be made on the basis of the angular momentum or other quantum numbers of the particle involved. If there is no need for an aristocracy among strongly interacting particles, may there not be democracy?

"Bootstrap" Dynamics

Composite particles owe their existence to the forces acting in channels with which the particles communicate. How do these forces arise and how can they be calculated?

The key concept behind the calculation is "crossing." Consider the following reaction involving four particles:

$$a + b \longleftrightarrow c + d$$

This says that the channel a,b is coupled to the channel c,d. The probability that this reaction will take place (in either direction) is expressed mathematically as the absolute value squared of the "reaction amplitude," which depends on the energies of the four particles involved. The principle of crossing states that the same reaction amplitude also applies to the two "crossed" reactions in which *ingoing* particles are replaced by *outgoing* antiparticles (indicated by a bar over a letter) thus:

$$a + \bar{c} \longleftrightarrow \bar{b} + d$$
$$a + \bar{d} \longleftrightarrow \bar{b} + c$$

These different reactions are distinguished by the signs of the energy variables, which are positive or negative according to whether ingoing or outgoing particles are involved, but if the reaction amplitude is known for any one of the three reactions, it can be obtained for the other two by inserting the proper signs for energy.

An example of crossing is the following pair of reactions involving neutrons and protons:

(a) $n + p \longleftrightarrow n + p$
(b) $n + \bar{n} \longleftrightarrow p + \bar{p}$

Both reactions are described by the same reaction amplitude, an important aspect of which can be depicted diagrammatically, as shown in the illustration on the preceding page. The first way of drawing the arrows in this diagram is appropriate to reaction a and the second to reaction b. The two figures differ, of course, only in the direction one reads them, either from bottom to top or left to right, as indicated by the arrowheads.

One interprets the first figure by saying that in a scattering collision between a neutron and a proton a meson is "exchanged," and it can be shown that this exchange is a way of representing the force acting between those two baryons. The interpretation of reaction b is that a meson that communicates with both the n,n and $p,\bar{p}$ channels provides a means for connecting the two channels of the reaction. Thus a single diagram corresponds both to a force in one reaction and to an intermediate particle in the crossed reaction. It follows that forces in a given channel may be said to arise, in general, from the exchange of intermediate particles that communicate with crossed channels.

With this as background we return to the idea mentioned in the introduction to this article, that the strongly interacting particles are all dynamical structures that owe their existence to the same forces through which they mutually interact. In short, the strongly interacting particles are the creatures of the strong interaction. We refer to this as the "bootstrap" hypothesis. It was formulated by one of the authors (Chew) and S. C. Frautschi at the University of California at Berkeley.

According to the bootstrap hypothesis each strongly interacting particle is assumed to be a bound state of those channels with which it communicates, owing its existence entirely to forces associated with the exchange of strongly interacting particles that communicate with crossed channels. Each of these latter particles in turn owes its existence to a set of forces to which the first particle makes a contribution. In other words, each particle helps to generate other particles, which in turn generate it. In this circular and violently nonlinear situation it is possible to imagine that no free, or arbitrary, variables appear (except for something to establish the energy scale) and that the only self-consistent set of particles is the one found in nature.

We remind the reader that in electromagnetic theory a few special particles (leptons and photon) are *not* treated as bound (or composite) states, the masses and coupling characteristics of each particle being freely adjustable. Conventional electrodynamics, as far as anyone knows, is not a bootstrap regime.

It is too soon to be sure that free variables are absent for strong interactions, but we shall close in an optimistic spirit by mentioning a fascinating possibility that would represent the ultimate contribution of the bootstrap hypothesis. If the system of strongly interacting particles is in fact self-determining through a dynamical mechanism, perhaps the special strong-interaction symmetries are not arbitrarily imposed from the outside, so to speak, but will emerge as necessary components of self-consistency. It is remarkable, and puzzling, that isotopic-spin symmetry, strangeness and now the broader eightfold-way symmetry have never been related to other physical symmetries. Perhaps their origin is destined to be understood at the same moment we understand the pattern of masses and spins for strongly interacting particles. Both this pattern and the puzzling symmetries may emerge together from the bootstrap dynamics.

II

CURRENT CONCEPTS: QUARKS AND QUANTUM CHROMODYNAMICS

5

Electron-Positron Annihilation and the New Particles

by Sidney D. Drell
June, 1975

Energetic collisions between electrons and positrons give rise to the unexpected particles discovered last November. They may help to elucidate the structure of more familiar particles

When matter and antimatter are brought together, they can annihilate each other to form a state of pure energy. A fundamental principle of physics demands that the reverse of that process also be possible: A state of pure energy can (quite literally) materialize to form particles of ponderable mass. When the matter and antimatter are an electron and a positron, the state formed by their annihilation consists of electromagnetic energy. It is a particularly simple state, since electromagnetism is described by a well-tested theory and is believed to be understood. For some time physicists have been eager to learn just what kinds of particles are created when an electron and a positron collide at high energy. During the past two years several experiments have provided a preliminary view of the annihilation process; the results have annihilated expectations. It is as the British poet Gerald Bullett described the arrival of spring:

> *Like a lovely woman late for her*
> *appointment*
> *She's suddenly here, taking us*
> *unawares,*
> *So beautifully annihilating*
> *expectation.*

The discoveries are the most startling and exciting to emerge from high-energy physics in a decade or more.

One reason for the great interest in these experiments is that they provide a means of testing a central concept of modern particle physics: the notion that the "herd" of supposedly elementary particles discovered during the past 25 years may actually be assemblages of only a few structureless entities that are truly fundamental. These constituent particles have been named quarks. Different versions of the quark theory make different predictions about what is to be expected in the aftermath of an electron-positron annihilation, and it was hoped that the experiments would help to determine which version is the correct one.

As it turned out, the results of an initial series of experiments in 1973 and early 1974 were not in accord with any of the predictions. Then, last November, as the measurements were being repeated and refined, two very massive particles were unexpectedly discovered. By coincidence the discovery of the first of the particles was announced simultaneously by physicists at two laboratories studying quite different reactions.

Four Forces

The existence of the new particles is in itself a surprise, but even more remarkable is their extraordinary stability. Although they decay to more familiar, less massive particles in a period that by conventional standards is very brief, their lifetime is about 1,000 times longer than that of other particles of comparable mass. This exceptional stability suggests that the new particles are fundamentally different from other kinds of matter. As yet their nature has not been satisfactorily explained, and their significance remains a subject of lively speculation. Theories abound and physics is in a state of great ferment, but we cannot be sure where the particles fit into the scheme of things.

Subatomic particles can be classified in broad families according to the kinds of interactions they participate in, or, as it is often put, according to the kinds of forces they "feel." The forces considered are the four fundamental ones that are believed to account for all observed interactions of matter: gravitation, electromagnetism, the strong force and the weak force.

In the everyday world gravitation is the most obvious of the four forces; it influences all matter, and the range over which it acts extends to infinity. For the infinitesimal masses involved in subatomic events, however, its effects are vanishingly small and can be ignored.

The electromagnetic force also has infinite range, but it acts only on matter that carries an electric charge or current. The photon is the quantum, or carrier, of the electromagnetic force, and when two particles interact electromagnetically, they can be considered to exchange a photon or photons. In the classification of particles the photon is in a category by itself: it has no mass and no charge and it does not participate in either strong or weak interactions.

The strength referred to in the names of the strong and the weak interactions is related to the rate at which the interactions take place. The strong force has a short range: its effects extend only about 10^{-13} centimeter, or approximately the diameter of a subatomic particle such as a proton. When two particles that feel the strong force approach to within this distance, the probability is very high that they will interact, that is, they will either be deflected or they will produce other particles. In contrast, particles that interact electromagnetically are 10,000 times less likely to interact under the same circumstances. If the strongly interacting particles pass each other at nearly the speed of light (3×10^{10} centimeters per second), then they must interact during the 10^{-23} second they are within range of each other. That is the characteristic time scale of the strong interactions.

Compared with the strong force, the weak force is feeble indeed: for collisions at low energy it is weaker by a factor of about 10^{13}. At higher collision energies the strength of the weak force increases, but even at the highest energies yet studied experimentally it is weaker than the strong force by a factor of about 10^{10}. Moreover, the range of the weak force is at most a hundredth that of the strong force. Two particles must approach to within 10^{-15} centimeter in order to feel the weak force, and even at that range the probability that they will interact is less than one in 10^{10}.

All particles except the photon are classified according to their response to these two forces. Those that feel the strong force are called hadrons; those that do not feel the strong force but do respond to the weak force are called leptons. Particles belonging to these two families have quite different properties.

The hadrons are subdivided into two classes called baryons and mesons. The baryons include the familiar proton and neutron (and it is the strong force that binds these particles together in atomic nuclei). The mesons include such particles as the pion and the kaon; they are generally less massive than the baryons, but they are all more massive than the leptons.

As a result of discoveries made over the past two decades the baryons and the mesons have become very large families of particles; there are in all more than 100 known hadrons, most of them massive and unstable. It was in an effort to explain this great proliferation of particles that the quark hypothesis was introduced independently in 1963 by Murray Gell-Mann and by George Zweig, both of the California Institute of Technology. The quark model posits that the unstable particles are excited states of the stable ones. Baryons are thought to consist of three quarks bound together and mesons to consist of a quark and an antiquark. Just as an atom can enter an

TWO-MILE-LONG ACCELERATOR at the Stanford Linear Accelerator Center (SLAC) was employed to generate high-energy electrons and positrons in the experiments that led to the discovery of the new particles. The main beam of the accelerator (extending under the highway) propels electrons into a target of tungsten, a third of the way down the accelerator. Some of the electrons collide with tungsten nuclei, creating electron-positron pairs. The positrons and the remaining electrons in the main beam are then further accelerated and injected into a storage ring, where the particles and antiparticles interact. The ring, called SPEAR, is to the right of the largest buildings. In it beams of electrons and positrons are confined in a single evacuated chamber. The beams circulate in opposite directions, guided by a magnetic field, and pass through each other twice each revolution. The two buildings straddling the ring enclose detectors that surround the regions where particle-antiparticle annihilations take place. The ring is about 250 feet in diameter.

		PARTICLE	SYMBOL	MASS	CHARGE	SPIN	LEPTON NUMBER	MU-NESS	BARYON NUMBER	LIFETIME
		PHOTON	γ	0	0	1	0	0	0	STABLE
	LEPTONS	ELECTRON	e^-	.5	-1	$\frac{1}{2}$	$+1$	0	0	STABLE
		POSITRON	e^+	.5	$+1$	$\frac{1}{2}$	-1	0	0	STABLE
		ELECTRON NEUTRINO	ν_e	0	0	$\frac{1}{2}$	$+1$	0	0	STABLE
		ELECTRON ANTINEUTRINO	$\bar{\nu}_e$	0	0	$\frac{1}{2}$	-1	0	0	STABLE
		MUON	μ^-	106	-1	$\frac{1}{2}$	$+1$	$+1$	0	10^{-6}
		ANTIMUON	μ^+	106	$+1$	$\frac{1}{2}$	-1	-1	0	10^{-6}
		MUON NEUTRINO	ν_μ	0	0	$\frac{1}{2}$	$+1$	$+1$	0	STABLE
		MUON ANTINEUTRINO	$\bar{\nu}_\mu$	0	0	$\frac{1}{2}$	-1	-1	0	STABLE
HADRONS	BARYONS	PROTON	p	939	$+1$	$\frac{1}{2}$	0	0	$+1$	STABLE
		ANTIPROTON	$\bar{p}$	939	-1	$\frac{1}{2}$	0	0	-1	STABLE
		NEUTRON	n	939	0	$\frac{1}{2}$	0	0	$+1$	10^3
		ANTINEUTRON	$\bar{n}$	939	0	$\frac{1}{2}$	0	0	-1	10^3
	MESONS	PION	π^+	137	$+1$	0	0	0	0	10^{-8}
			π^-	137	-1	0	0	0	0	10^{-8}
			π^0	137	0	0	0	0	0	10^{-15}
		RHO MESON	ρ^+	750	$+1$	1	0	0	0	10^{-23}
			ρ^-	750	-1	1	0	0	0	10^{-23}
			ρ^0	750	0	1	0	0	0	10^{-23}
		PSI (3095)	ψ	3095	0	1	0	0	0	10^{-20}
		PSI (3684)	ψ	3684	0	1	0	0	0	10^{-20}

SUBATOMIC PARTICLES are classified according to the kinds of interactions in which they participate. The hadrons take part in "strong" interactions; the leptons do not; the photon interacts only electromagnetically. The hadrons are divided into mesons and baryons, which differ in their spin angular momentum and in other properties. The classification is reflected in quantum numbers such as lepton number, mu-ness and baryon number. The newly discovered particles, psi(3095) and psi(3684), are mesons. Their most perplexing property is their lifetime, which is 1,000 times longer than that of other particles of comparable mass, such as the rho meson.

excited state when the orbital configuration of its electrons is changed, so a hadron, by analogy, enters an excited state of higher energy (or mass) when the configuration of its constituent quarks is altered.

A further insight into the nature of hadrons was provided in the late 1960's by experiments in which high-energy electrons were scattered by protons and neutrons. The experiments were performed at the Stanford Linear Accelerator Center (SLAC) by a group of investigators from the Massachusetts Institute of Technology and SLAC under the direction of Jerome I. Friedman, Henry W. Kendall and Richard E. Taylor. Once again the results inspired an analogy with events on the atomic scale. In 1911 Ernest Rutherford investigated the scattering of alpha particles by atoms and was led to predict the existence of a massive nucleus within the atom. Similarly, the SLAC experiments revealed an internal structure within the individual hadron; the pattern of electron scattering suggested the existence of pointlike substructures, which were named partons. Partons and quarks are believed to be equivalent [see "The Structure of the Proton and the Neutron," by Henry W. Kendall and Wolfgang K. H. Panofsky; SCIENTIFIC AMERICAN, June, 1971].

The leptons are a much smaller class of particles than the hadrons. There are just four: the electron and its neutrino and the muon and its neutrino (and their four antiparticles). The electron has a mass of about .5 MeV (million electron

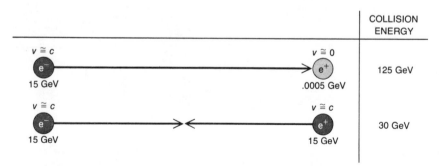

COLLISION ENERGY in electron-positron annihilations depends on the initial energy and momentum of both particles. When one particle is stationary and the other has a velocity near c (the speed of light in vacuum), the collision energy is only a small fraction of the energy expended in accelerating the moving particle. When two particles collide with equal and opposite velocity, all their energy is made available for the creation of new particles.

volts) and an electric charge of −1 unit. The muon has the same charge, but it is 207 times as massive as the electron. Both kinds of neutrino are without mass and charge. Because the electron and the muon are charged they can interact electromagnetically as well as by the weak force; the neutrinos feel only the weak force.

Among the leptons no spectrum of excited states comparable to that of the hadrons has been discovered. Nor have scattering experiments yielded any suggestion of an internal structure. The leptons are thus fundamentally different from the hadrons. They are not composed of quarks or partons but are themselves apparently pointlike. It is possible they are analogues of the pointlike constituents of hadrons.

Quantum Numbers

Physicists identify and describe subatomic particles by assigning them quantum numbers. Each number designates a property that is conserved, or left unchanged, when particles interact. Some quantum numbers, such as electric charge and spin angular momentum, refer to physical, measurable attributes of the particle. Others are more abstract; they denote family resemblances among particles and provide a valuable bookkeeping system for classifying particles and their interactions in algebraic form. Baryons and mesons, for example, are distinguished by a property called baryon number. Baryons are assigned a value of +1, antibaryons −1 and mesons 0; to say that baryon number is conserved is merely to say that baryons never turn into mesons. The least massive baryon, the proton, cannot decay into the less massive mesons or leptons because in such a process baryon number would be changed. The proton is therefore stable.

There are also conserved properties, or quantum numbers, of the leptons that ensure the stability of those particles. Lepton number is a quantity handled in the same way as baryon number. Its conservation prohibits the transformation of individual leptons into pure energy or into hadrons, just as hadrons are forbidden to become leptons. For instance, a positron has the same charge as a proton, and it can be accelerated until its mass is equal to that of a proton, but no positron has ever been observed to change into a proton.

In the same way a quantum number named mu-ness divides the leptons into two groups. The muon and the muon neutrino have mu-ness of +1, their anti-

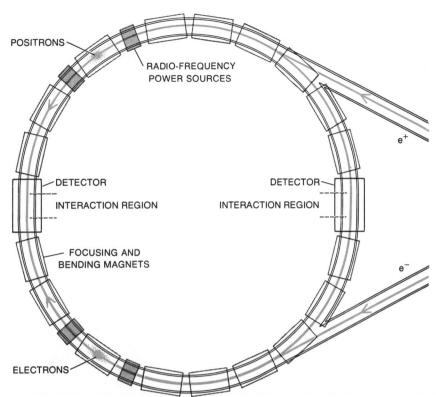

COUNTERROTATING BEAMS of electrons and positrons in the SPEAR storage ring are each confined to a small "bunch" a few centimeters long. Their trajectories are determined by the magnetic field and their energy is maintained by the input of radio-frequency power; a single field guides both electrons and positrons, since the oppositely charged particles respond oppositely to it. Moving at nearly the speed of light, the beams pass through each other several hundred thousand times per second. Each beam can have an energy of up to 4 GeV (billion electron volts), so that total collision energies of up to 8 GeV can be achieved.

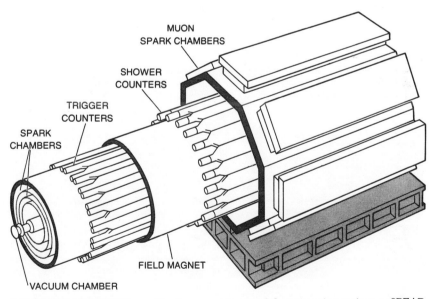

MAGNETIC DETECTOR completely surrounds one of the interaction regions at SPEAR. It consists of several layers of scintillation counters and spark chambers, which are here extended along their axis for clarity. Both kinds of detector produce an electrical impulse when a charged particle passes through them. Neutral particles cannot be detected. Information from the detectors is recorded electronically and employed to identify the particles, to reconstruct their trajectories and to determine which events are the result of an electronpositron annihilation and which are derived from other, extraneous sources. The momentum of a particle is determined by measuring its deflection in magnetic field of detector.

particles have mu-ness of −1 and the electron, the positron, the electron neutrino and the electron antineutrino all have zero mu-ness. An electron can transfer its mass and charge to more massive particles and become an electron neutrino, but because mu-ness is conserved it can never change simply into a muon or a muon neutrino.

Baryon number, lepton number and mu-ness are always conserved; certain other quantum numbers, however, are conserved only in some interactions. They are said to describe "approximate" conservation laws. Several properties, for example, are conserved in strong interactions but not in weak or electromagnetic ones.

In addition to the several conserved quantum numbers, energy and momentum are always conserved. Energy and momentum can be transferred from one particle to another, but the sum of the energies and momenta before an interaction must be exactly equal to the sum afterward.

In the annihilation of matter and antimatter the arithmetic of quantum numbers is particularly simple. The concept of antimatter was introduced in 1928 by P. A. M. Dirac, and it is now a firmly established principle that for every particle there exists an antiparticle. All the quantum numbers of an antiparticle are "opposite" those of the corresponding particle, that is, the sum of the quantum numbers of the particle and antiparticle is zero. The electron has lepton number +1, mu-ness 0 and charge −1; for the positron the corresponding values are

−1, 0 and +1. For each quantum number the sum is zero.

When matter and antimatter meet, their quantum numbers, being opposite, simply cancel. Properties such as charge and lepton number are conserved, but they are also eliminated! Furthermore, from the state of energy thus created, particles having virtually any combination of quantum numbers can, at least in principle, be formed, as long as the sum of the quantum numbers of the products remains zero. In particular, a pair or more than one pair of particles and antiparticles can be created. There is only one constraint on the process: the energies of a colliding particle and antiparticle do not cancel, and in the annihilation energy must be conserved just as it would be in any other interaction. Momentum must also be conserved.

From these considerations it can be seen that some processes that are forbidden in interactions of ordinary particles are permissible in the annihilation of particle-antiparticle pairs. For example, although an electron is prohibited by the conservation of mu-ness from becoming a muon (except with the emission of an electron neutrino and a muon antineutrino, a rare process), it is entirely possible for an electron and a positron to annihilate each other and for a muon and an antimuon to materialize from the energy state formed. It is necessary only that momentum be conserved and that the initial particles have sufficient energy to account for the rest mass of the muon-antimuon pair [see "Electron-Positron Collisions," by Alan M. Litke and Rich-

ard Wilson; SCIENTIFIC AMERICAN, October, 1973].

It is also possible for the annihilation of an electron-positron pair to give rise to hadrons. The reaction is written $e^-e^+ \rightarrow$ hadrons, where e^-e^+ represents the annihilating pair. It was during an investigation of this process at SLAC that the new particles were discovered.

The particles were found by a group of 35 physicists from the Lawrence Berkeley Laboratory and SLAC, led by Burton Richter, William Chinowsky, Gerson Goldhaber, Martin L. Perl and George H. Trilling. The less massive of the two particles was discovered simultaneously at the Brookhaven National Laboratory in experiments investigating a process that is the inverse of the reaction studied at SLAC. The Brookhaven experiments were performed by Samuel C. C. Ting and his colleagues from M.I.T. and Brookhaven; they studied the production of electron-positron pairs in collisions of hadrons. The detection of the particle by Ting's group was a technical tour de force: it is produced by their technique only once in 10^8 events.

Particle Storage Rings

Because hadrons are much more massive than electrons, hadrons can be created only in annihilations at high energy. In order to achieve the required collision energies it has been necessary to build machines known as particle storage rings. In these rings particles and antiparticles are made to circulate and then to collide head on [see "Particle Storage Rings," by Gerard K. O'Neill; SCIENTIFIC AMERICAN, Offprint 323].

Ordinary particle accelerators can produce electrons or positrons of very high energy, but when such a particle strikes an electron in a stationary target, little of that energy is liberated in the collision. This effect is a consequence of the special theory of relativity, which states that when a particle acquires a large total energy, it also acquires a large total mass (according to the equation $E = mc^2$, where c is the speed of light in vacuum). Once an electron or a positron has been accelerated to about 1 GeV (billion electron volts) its velocity is within a few centimeters per second of the speed of light. Any additional energy imparted to it has negligible effect on its velocity and goes almost entirely toward increasing its mass. The two-mile linear accelerator at SLAC can produce positrons with an energy, or mass, of 15 GeV. The collision of such a particle with an electron at rest (which therefore has a mass of about .0005 GeV, 30,000 times

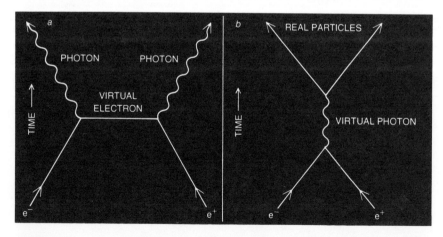

ANNIHILATION of an electron and a positron produces electromagnetic energy, or photons. The diagram depicting the event is called a Feynman graph (after Richard P. Feynman); distance is represented on one axis and time on the other. If momentum is to be conserved, two photons must be emitted (a); the photons, like the initial particles, have equal and opposite momentum and thus net momentum of zero. When only one photon is created (b), it must have large energy but zero momentum, a condition that is impossible for a real photon. The particle formed is called a virtual photon; it can never be detected and it quickly decays into real particles with zero total momentum. The production of two photons is said to take place through the exchange of another virtual particle: a virtual electron.

smaller than the mass of the positron) is like the collision of a charging elephant with a mouse. The mouse may bounce back or it may be crushed, but in either case the elephant's kinetic energy will be little changed by the collision. In the case of the stationary electron and the 15-GeV positron only about .125 GeV would be made available to create new particles. That is less than 1 percent of the energy supplied to the accelerated particle, and it is too little to produce even a single pion, the least massive hadron.

The result is quite different when an electron and a positron moving with equal velocity in opposite directions collide head on. In this case the sum of the momenta of the two particles is zero, and all their energy appears in the products of the annihilation. The collision is analogous to the head-on encounter of two charging elephants. In the storage-ring facility at SLAC, which is called SPEAR, electrons and positrons can be made to collide with energies of up to about 4 GeV each, for a total energy of 8 GeV, enough to generate a rich spectrum of hadrons. To produce such collision energies with stationary targets would call for a particle beam having an energy of 64,000 GeV; the accelerator required would be some 6,000 miles long.

The SPEAR storage ring was built under the direction of John R. Rees and Richter. In it "bunches" of electrons and positrons circulate in opposite directions in a single toroidal chamber about 250 feet in diameter. They are confined by a magnetic field, and their energy is maintained by the input of radio-frequency energy. The same magnetic field sustains both counterrotating bunches; because the particles are oppositely charged they react oppositely to it [see top illustration on page 61].

The beams collide at two regions on the perimeter of the ring, where detectors have been installed to record and analyze the products of interactions. The probability of even a single e^-e^+ annihilation in any one pass-through of the bunches is quite low, but because the particles' velocity is nearly that of light they pass through one another several hundred thousand times per second, so that an acceptable reaction rate is achieved. The beams of particles can be stored in the ring for minutes at a time.

Virtual Particles

What happens when an electron and a positron meet and annihilate each other with a combined energy of a few billion electron volts? Because the particles

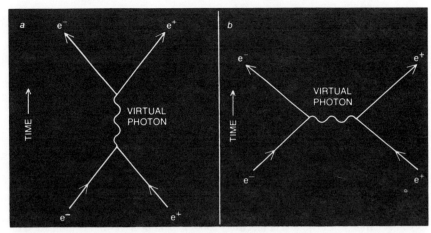

TWO INTERPRETATIONS are possible when the product of an encounter between an electron and a positron is another electron-positron pair. The initial pair may have annihilated (a), producing a virtual photon that materialized into an electron and a positron. Or the particles may have passed nearby and been scattered by the exchange of a photon (b).

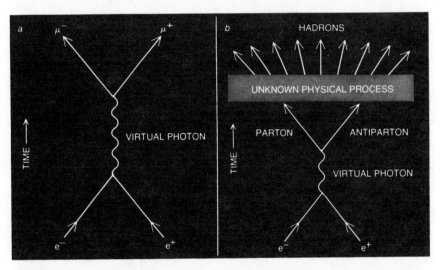

MUONS AND HADRONS can be produced in electron-positron encounters only by the annihilation of the particles, not by scattering. The creation of muons (a) requires that the quantum number mu-ness be changed for each of the particles, although the total mu-ness of the pair remains zero. Hadrons are thought to be formed (b) through the creation of a parton and an antiparton (or a quark and an antiquark). The partons interact to form hadrons.

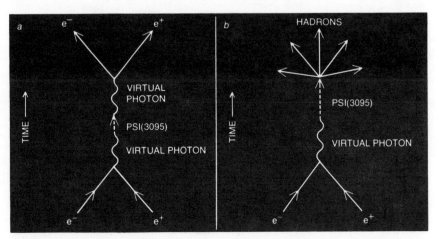

NEW PARTICLES, psi(3095) and psi(3684), materialize from the virtual photon when it has exactly the right energy (3.095 GeV and 3.684 GeV respectively). They can decay through a virtual photon into leptons (a), such as an electron and a positron, or into hadrons (b). The hadrons most commonly produced by psi particles are pions, the least massive of the hadrons.

are leptons they do not feel the strong force, and at the energies studied so far the weak interactions are feeble enough to be neglected. The particles are electrically charged, however, so that they do feel the electromagnetic force, and the energy produced by their annihilation is (to a very good approximation) entirely electromagnetic. In other words, the electron and the positron annihilate each other, canceling their electric charges and lepton numbers, to produce a very energetic photon (a gamma ray).

The photon emitted is not, however, a "real" photon such as those that are observed in nature as the quanta of electromagnetic energy. It cannot be real because it has the wrong proportions of energy and momentum, quantities that must be conserved in all interactions. For the photon, which has no mass and which travels at the speed of light, the relation of momentum to energy is constant: the momentum is a fixed fraction of the energy, equal to the energy divided by c. This energy-momentum relation cannot be reconciled with the energy and momentum of the colliding particles. In the storage ring the electron and positron move with equal energy but opposite momentum, and the state formed by their annihilation must therefore have large energy but zero momentum. A photon cannot have that combination of properties.

One possible resolution of this dilemma is for the annihilation to produce two photons that have equal but opposite momenta, thereby satisfying the condition that the sum of the momenta of the products be zero. This reaction does in fact take place, and measurement of it is of major interest. Generally, however, the annihilation process generates as few photons as it possibly can. The probability that an electron or a positron will interact with or produce a single photon is measured by a dimensionless number called the fine-structure constant, equal to about $1/137$. For each additional photon the probability is reduced by a higher power of the same factor.

The most likely outcome of the annihilation is therefore the creation of a single photon. As we have seen, however, it cannot be a real particle; it is called a virtual photon, and its most important characteristic is that it can never be observed; it can never emerge from the reaction as radiation. The virtual photon serves merely as a coupling between the initial electron-positron pair having zero total momentum and some subsequent ensemble of particles that must also have zero total momentum.

The virtual photon cannot be observed because it decays before it can be detected. According to the uncertainty principle formulated by Werner Heisenberg, the lifetime of a virtual particle is necessarily too brief for the particle to be observed. In the case of the recent electron-positron annihilation experiments the virtual photon materializes in less than 10^{-25} second into particles with the correct combination of energy and momentum.

Particle Production

When the virtual photon decays, several kinds of particles can be created. At the energies investigated so far pairs of electrons and positrons, pairs of muons and antimuons, and hadrons have all been observed.

If the collision yields an electron-positron pair, the annihilation and rebirth of such a pair is phenomenologically indistinguishable from the mere elastic scattering of the incident electron and positron. In the first case the pair disappears in forming a virtual photon and then reappears; in the second the two particles are said to exchange a photon and are deflected, so that their direction is changed but the magnitudes of their energy and momentum are not [see top illustration on preceding page].

The creation of a muon-antimuon pair is not complicated by this ambiguity.

RESONANCE detected in electron-positron annihilations at SPEAR signifies the presence of a new particle, psi(3095). It was discovered last November by a group of physicists under the direction of Burton Richter, William Chinowsky, Gerson Goldhaber, Martin L. Perl and George H. Trilling. The resonance represents a greatly increased probability of interaction between the colliding particles at the resonance energy. In this case the electrons and positrons are about 150 times as likely to annihilate each other and yield hadrons when their combined energy is 3.095 GeV as they are at adjacent, "background" energies. The psi(3095) resonance is exceptionally narrow, which indicates that the lifetime of the particle is long.

Because mu-ness must be conserved the pair can be created only through the annihilation of the electron and the positron. It is also necessary, of course, that the e^-e^+ pair have enough energy to account for the mass of the muons. These reactions are of great interest because they provide a method of testing the validity of present theories of electromagnetism at high energy. In addition there is enormous interest in studying the reactions that lead to the production of hadrons.

Processes that involve only leptons and photons can be described entirely in terms of the electromagnetic interaction. To predict hadron production, however, one must have a theory of the structure of hadrons; as yet there is no completely satisfactory theory. Most ideas of hadron structure that are under consideration today rely on the concept of partons or quarks as constituents of the hadron. They predict that at high energy a virtual photon can decay into a parton and an antiparton (or a quark and an antiquark). The parton-antiparton pair is then transformed by the strong interaction into hadrons that emerge in the aftermath of the collision.

Even though partons and quarks have not been observed in isolation, many of their properties are described in detail by theory; they are *required* to have certain properties in order to explain the properties of the families of hadrons that have been observed. In addition the interpretation of the scattering experiments performed at SLAC (which originally led to the parton hypothesis) indicates that at high energy quarks or partons behave as independent, pointlike entities, just as the leptons do. For this reason it was expected that once the total energy of the colliding particles exceeded a threshold value, suggested by the scattering experiments to be about 2 GeV, the production of leptons and partons would obey similar rules. In particular it was predicted that above 2 GeV the probability of producing hadrons would vary with the collision energy in the same way that the probability of producing a pair of muons varies. This relation was expressed mathematically by stating that the ratio of hadrons to muon-antimuon pairs would be constant and independent of the collision energy. Before the experiments were performed there was argument over what the numerical value of this ratio would be, but there was little doubt that the ratio would in fact be constant at all energies within reach of experiment.

The disagreement arose because sev-

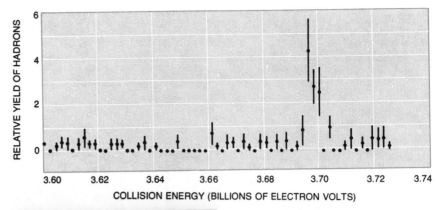

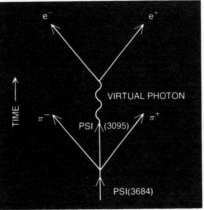

PSI(3684) RESONANCE is neither as tall nor as narrow as that of the psi(3095) particle. The graph (*above*) is at lower resolution than the one on the opposite page, that is, the yield of the electron-positron collisions was measured at fewer energies. The yield is expressed in relative numbers of hadrons produced at each energy. Psi(3684) decays spontaneously into psi(3095) with the emission of a pair of charged pions (*left*). Psi(3095) can then decay in turn, producing other particles, such as an electron-positron pair or hadrons. Because psi(3684) can produce psi(3095) by decay, it has been suggested that the more massive particle is an excited state of the less massive one.

eral versions of the quark theory predict different values for the hadron/muon ratio. In each case the value is calculated simply by adding the squares of the charges of all the quarks postulated by the model [*see lower illustration on page 69*]. The original formulation of the quark hypothesis, for example, predicts a value of 2/3. Three of the more prominent variations on the theory give values of 2, 10/3 and 4.

Initial measurements of the ratio were made in 1973 at Frascati in Italy and at the Cambridge Electron Accelerator in Cambridge, Mass.; they were soon followed by the first round of measurements with the SPEAR storage ring. To the surprise of most particle physicists, none of the predictions was confirmed; the experiments indicated that the hadron/muon ratio is not constant at all. At 2 GeV the value appeared to be about 2, and it increased gradually to about 5 at 5 GeV. In other words, collisions at 2 GeV were twice as likely to produce hadrons as to produce muon pairs, and at 5 GeV hadrons were about five times as likely. This disturbing development had not yet been accommodated by theory (and it still has not been) when the unexpected massive particles were encountered last November.

At the SPEAR storage ring the first of the new particles was discovered dur-

ing a second round of measurements of the hadron/muon ratio made at many (and more closely spaced) values of the collision energy. The mass of the particle has now been determined accurately as being 3.095 GeV, and the particle has been named by the SPEAR group psi(3095). (The particle was given another name, *J*, at Brookhaven, but here I am adopting the SPEAR nomenclature.) A second, heavier particle was subsequently found at SPEAR, and it was designated psi(3684), to denote its mass of 3.684 GeV. The heavier particle can decay to form the lighter one, along with two pions. The Brookhaven workers have searched for the 3.684-GeV resonance, but they have found that it cannot be detected by their methods, a fact that may provide a clue to the properties of the particle.

The New Particles

The new particles are unstable, and like all other very massive and short-lived particles they are detected as "resonances," or enhancements of the probability of an interaction [see "Resonance Particles," by R. D. Hill; SCIENTIFIC AMERICAN, Offprint 290.] As the energy of the colliding beams is increased in small steps a sudden peak in the production of particles is observed at the resonance energy; when the beam energy is increased further, the production drops off again. Such a pattern indicates that a particle exists with a mass equal to the combined masses of the colliding particles; when the colliding particles have exactly the required ener-

gy, they are more likely to interact than they are at other energies. In the case of psi(3095) the probability that the electron and the positron will interact was observed to increase by a factor of about 150 at the resonance energy compared with its value at adjacent energies.

The height of the resonance peaks is related to their most remarkable feature—their narrowness. The width of a resonance represents the uncertainty in the determination of the energy at which the resonance occurs. This uncertainty in the energy is in turn related to the lifetime of the particle by Heisenberg's uncertainty principle. The equation that gives the lifetime is $\Delta E \Delta t = h/2\pi$, where ΔE is a measure of the uncertainty in the determination of the energy, Δt is a measure of the lifetime and h is Planck's constant (approximately 6.6×10^{-27}). Only for a stable particle—one that has an indefinitely long lifetime, as the proton apparently does—can the energy or mass be precisely defined. In that case Δt is large and ΔE is correspondingly small; the lifetime is unlimited and the resonance width is vanishingly small, so that the energy of the particle can be known with arbitrarily great precision.

The width of the psi(3095) resonance has been determined to be about 77 KeV (thousand electron volts), which represents a very sharp peak. Substituting this value in the equation above yields a lifetime of about 10^{-20} second. The psi(3684) resonance is somewhat broader, and that particle therefore decays faster, with a lifetime of about 10^{-21} second.

Although 10^{-20} second is an almost unimaginably brief interval, far too brief to be measured directly, it is 1,000 times longer than the expected lifetime of such a particle. All other heavy resonances are far broader (their width is typically measured in millions of electron volts rather than thousands), and the particles they represent decay in the characteristic time of the strong interaction: about 10^{-23} second. We are immediately compelled to ask what properties of psi(3095) hold it together so long. Does it exhibit a new kind of structure? Is there a previously unknown quantum number that nature wants to conserve but that must be changed when psi(3095) decays? Questions of this kind are among the most fundamental that can be asked in the physics of elementary particles.

Although the psi particles are themselves mystifying, their discovery has to some extent simplified, if not clarified, the earlier measurements of the hadron/muon ratio. It now appears that for energies between 2 GeV and 3.8 GeV the ratio is roughly constant at about 2.5; the only significant deviations are those associated with the two psi resonances. At higher energy the ratio increases, reaches a broad peak at about 4.1 GeV,

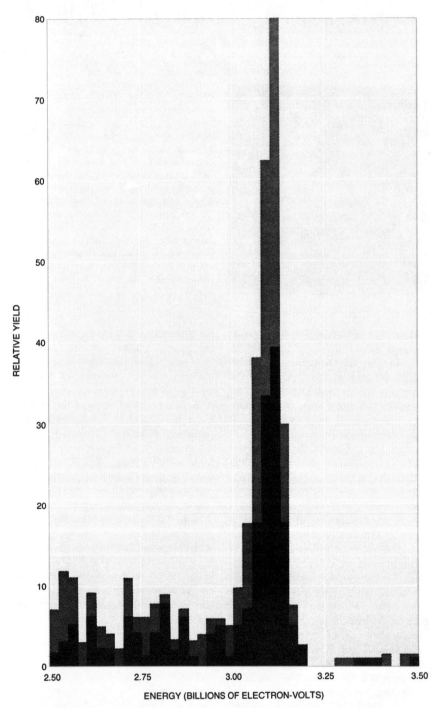

RELATIVE YIELD

ENERGY (BILLIONS OF ELECTRON-VOLTS)

HADRON COLLISIONS also result in the production of one of the new particles. The resonance is detected as an increased probability of the production of electron-positron pairs in the collision of protons with beryllium nuclei; this process is the inverse of an electron-positron annihilation. The production of the new particle in hadron collisions was observed by a group of physicists, directed by Samuel C. C. Ting, working at the Brookhaven National Laboratory. They discovered it simultaneously with the SPEAR group and named it *J*. The psi(3684) particle has proved to be undetectable in the reaction studied at Brookhaven. The 3.095-GeV resonance is measured by the number of electron-positron pairs detected. Measurements at two calibrations of the detector are superposed (*gray and black*).

then stabilizes at a value of about 5 [*see illustration at right*]. Recent experiments at SPEAR have confirmed that the hadron/muon ratio remains approximately constant up to 6 GeV. The causes of the observed peculiarities near 4.1 GeV are not yet understood. The broad peak may or may not represent an additional resonance or perhaps several resonances.

Quantum Electrodynamics

The interpretation of the puzzling events revealed by electron-positron collisions is made substantially easier by the fact that the initial state emerging from the annihilation—the virtual photon—is simple and well understood. Because it is formed in a purely electromagnetic process it is described by the theory of quantum electrodynamics, one of the most successful theories in physics.

Quantum electrodynamics is the theory constructed by imposing the laws of quantum mechanics on the classical theory of electromagnetism. It is a basic tenet of quantum electrodynamics that electrons, when they interact with electromagnetic radiation, act as point charges. This assumption, however, may be only an idealization that, although useful at large distances, fails to describe events at subnuclear dimensions.

Physicists are eager to test the theory of quantum electrodynamics at ever higher energy and with greater precision in order to probe it in finer detail. It is particularly important to know whether or not quantum electrodynamics is valid for the very-high-energy collisions studied at SPEAR, since the theory is central to the interpretation of hadron structure.

Quantum electrodynamics has been tested at distances ranging from more than 250,000 miles (in measurements of the earth's magnetic field) to subatomic dimensions much smaller than 10^{-8} centimeter (in describing the detailed spectrum of the hydrogen atom). The atomic measurements have been made so precisely that theory and experiment agree to roughly four parts in 10^9.

In interactions of electrons and positrons whose final products are either electron-positron pairs, muon-antimuon pairs or two or more real photons only the electromagnetic interaction makes a significant contribution. The entire process, therefore, should be comprehended by quantum electrodynamics and can be employed to test the theory. Another application of the uncertainty principle demonstrates that at the energies available with existing storage rings such events probe distances as small as 10^{-15}

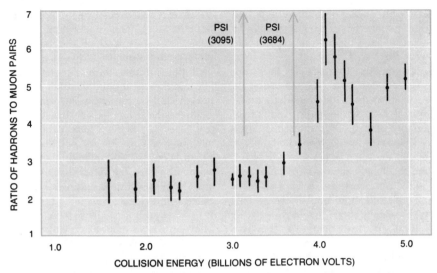

RATIO of hadrons to muon-antimuon pairs produced in electron-positron annihilations was predicted to be constant at collision energies above 2 GeV. The SPEAR experiments demonstrated that the ratio is roughly constant up to about 3.8 GeV if the effects introduced by the two psi particles are ignored. It rises to a peak at about 4.1 GeV, then declines again to a nearly constant value of approximately 5. Various versions of the quark model predict different constant values for the ratio, but none of them explains the observed fluctuations.

centimeter, or roughly 1 percent of the diameter of a hadron. Even at that small scale experiment has so far revealed no flaws in the theory.

The quantitative verification of quantum electrodynamics is the 20th-century parallel to the exciting voyage of Isaac Newton 300 years ago. The elegant simplicity of nature was first comprehended when Newton realized that the same law of gravity that governs the apple's fall to the ground also describes the motion of the planets, hundreds of millions of miles apart. We have now also learned that nature applies the same laws of electrodynamics both on a large scale, extending to about 80 earth radii, and in a realm less than a millionth the size of an atom. Quantum electrodynamics has been verified for distances encompassing more than 25 orders of magnitude.

If a colliding electron-positron pair can create a psi(3095) resonance, then of necessity the resonance particle, by the reverse process, must be able to decay into an electron-positron pair. (Indeed, that is how the particle was discovered by Ting and his colleagues at Brookhaven.) Psi(3095) decays in this way about 7 percent of the time, and muon pairs are produced in another 7 percent of the decay events. The remaining 86 percent of the time the resonance decays to yield hadrons. Since both muon pairs and electron pairs are created through purely electromagnetic forces, quantum electrodynamics should describe these processes, and it is possible to determine to what extent the two mechanisms interfere with each other.

In practice the interference is detected by measuring the ratio of muon pairs to electron pairs produced at energies near the psi(3095) resonance. Of particular importance is the observed decline in the ratio just below the peak energy of the psi(3095) resonance. On fundamental theoretical grounds such an "interference dip" is expected only when the electron-positron annihilation has two possible final states and the two states are, with regard to the electromagnetic interactions, interchangeable. In this instance the two final states can only be the photon and psi(3095), and the experiment has the important implication that psi(3095) has the same quantum numbers as the photon.

This discovery enables us to specify some of the characteristics of the new particles. For example, psi(3095), like the photon, must have a spin angular momentum of 1. Thus our detailed understanding of quantum electrodynamics provides us with information on the new particles.

Color and Charm

There is more to the psi particles, however, than their quantum numbers alone. In particular we want to know what kind of quarks they are made up of. For now all attempts to deduce their quark constitution must be considered speculations only, because experiment has not provided the evidence to discriminate conclusively among a welter of competing theories; nevertheless, two proposals that have been given considerable attention deserve discussion. Both were proposed to explain unrelated phenomena long before the psi particles were discovered.

Quarks, like the particles they compose, are assigned quantum numbers. All of them have spin angular momentum of 1/2, for example, and baryon number of 1/3. Of the original triplet of quarks proposed by Gell-Mann and Zweig, the u quark has a charge of 2/3 and the d and s quarks have a charge of −1/3. Antiquarks, of course, have the opposite quantum numbers. According to these assignments, the baryons, being made up of three quarks, must have a half-integral spin, a baryon number of +1 and a charge of +2, +1, 0 or −1. The mesons, as aggregates of a quark and an antiquark, must have an integral spin, a baryon number of 0 and a charge of +1, 0 or −1.

This ingenious scheme neatly accounted for all the particles and resonances that had been observed when it was proposed, and it soon proved its predictive power by postulating an unknown resonance that was promptly discovered. It contains a deeply disturbing peculiarity, however: the quarks are required to be particles with a half-integral spin but they do not behave as such particles are expected to.

All observed particles with a spin of 1/2 obey Wolfgang Pauli's exclusion principle, which demands that no two be in an identical state. The electrons of an atom, for example, always differ in at least one quantum number; if they have the same orbital configuration, they have opposite spin. Our understanding of atomic structure and of the periodic table of the elements is based on this concept. Particles with integral spin (such as the mesons and the photon) are not affected by the exclusion principle; arbitrarily large numbers of them can be assembled in the same state. (The half-integral-spin particles are said to obey Fermi-Dirac statistics, and they are called fermions; the integral-spin particles obey Bose-Einstein statistics and are called bosons.)

Quarks seem to violate these rules. They must have spin of 1/2, but on the other hand in constructing the baryons it is necessary to assume that two or more are bound together in the same state. The paradox threatens to do violence to some basic and cherished theoretical principles. It can be resolved, however, simply by insisting that quarks do obey the exclusion principle. All that is necessary to make them conform to the rule is to endow them with a new quantum number having three possible values, so that the quarks bound together in a baryon, although identical in all other properties, can differ in this new one. The new property, first suggested in 1964 by Oscar W. Greenberg of the University of Maryland, is called color, although it has nothing to do with vision or the color of objects in the macroscopic world; in this context color is merely a label for a property that expands the original ensemble of three quarks to nine. Each quark of the original triplet can appear in any of three colors—say red, yellow or blue.

An incidental feature of the color hypothesis is that the addition of six more quarks to the original three enables us to reformulate the quark theory with integral charges rather than fractional ones. A model of this kind was constructed by Moo-Young Han of Duke University and Yoichiro Nambu of the University of Chicago in 1965.

All versions of the color theory assume that in the known baryons the three colors of quarks are equally represented; as a result the particle exhibits no net color. Similarly, the mesons are made up of equal proportions of red, yellow and blue quark-antiquark pairs and are also colorless. Indeed, some physicists have speculated that in nature all particles may be colorless. One of a class of proposed interpretations of the psi particles suggests that they may be the first observed states of colored matter.

The second important theory that has been invoked to explain the psi particles is called the charm hypothesis. It was proposed by a number of theorists in 1964 simply for the sake of symmetry. (In particle physics that is not a trivial motive, since every symmetry has an associated conservation law.) In order to construct a parallel to the four known leptons, the theory adds to the three original quarks a fourth, designated c for charm. The c quark has a charge of 2/3, and it has +1 unit of the new quantum number charm; all the other quarks have zero charm.

In 1970 the new quark and its quantum number were given an important role in physics through the work of Sheldon L. Glashow, John Iliopoulos and Luciano Maiani of Harvard University. They invoked the charm quark in order to explain the suppression of certain particle decays that in the three-quark model should have proceeded normally.

Significance of the Particles

If the c quark exists, of course, one would expect to observe charmed states of matter, such as a baryon made up of a c quark and two other quarks. No charmed particles have been observed, and hence it is believed that charmed quarks are much more massive than uncharmed ones. The psi particles are not believed to be charmed either, since they are produced electromagnetically and the electromagnetic interactions conserve charm, but it has been suggested that they could be mesons consisting of a c quark and an anti-c-quark. That com-

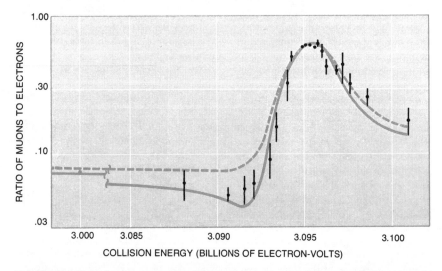

INTERFERENCE between entirely electromagnetic events and events that involve the strong interaction is detected by measuring the ratio of muon pairs to electron pairs produced in annihilation experiments. The presence of interference is an indication that psi-(3095) has the same quantum numbers as the photon. If the quantum numbers are the same, theory predicts that the muon/electron ratio will be depressed at energies just below the resonance energy (*solid colored line*); if the numbers are different, there will be no interference (*broken colored line*). Experimental results (*black dots and bars*) show interference.

bination would not exhibit charm, because the charm quantum numbers of the two quarks would cancel.

The first requirement of any theory of the new particles is that it explain their anomalously long lifetime. At the moment neither color nor charm seems to offer an entirely satisfactory explanation.

If psi(3095) and psi(3684) are considered to be colored particles, then it is assumed that they live long because there are no other particles of lower mass to which they can transfer their color when they decay. In the strong interactions color is conserved, so that a colored particle could not decay by the strong force. A particle might well be able to change color, however, in the electromagnetic interactions; the unit of color would be carried off by a photon, leaving decay products of colorless hadrons. As we have seen, the emission of a photon is suppressed by a factor of 1/137, and so the requirement that psi(3095) decay electromagnetically offers a partial explanation of its stability. The actual suppression of the decay involves a factor of about 1,000 instead of 137, but the discrepancy might be accounted for.

The hypothesis that the psi particles are colored is subject to experimental test by at least two methods. First, the gamma rays needed to carry off color in the electromagnetic decay should be detected among the decay products of the particles. Second, if the psi particles do represent colored particles, they cannot be the only ones; all versions of the color theory predict families of associated colored mesons, some of which would be electrically charged.

The charmed-quark theory is simpler and therefore more appealing, but it offers no compelling explanation for the psi lifetime. There is no known mechanism that would prevent a meson made of a charmed quark and antiquark from simply transforming itself into an ordinary quark and antiquark. In this process there are no quantum numbers to be conserved, since in the particle-antiparticle pair charm and all other properties cancel. The only apparent solution is to postulate a new law of nature, to declare arbitrarily that the decay of mesons made of a charmed quark and antiquark is inhibited by a factor of 1,000. There is precedent for such a rule in the theoretical treatment of other meson decays, but psi(3095) involves a larger inhibition factor. Moreover, it is a mystery why such rules should be required at all.

In spite of this weakness the charm hypothesis has attractive elements. The existence of resonances corresponding to a charmed quark and antiquark meson was predicted before the psi particles were discovered. The existence of the particle was discussed by Thomas W. Appelquist and H. David Politzer of Harvard, who named the hypothetical entity charmonium. They also suggested that it could be formed in electron-positron annihilations.

The charm hypothesis is also susceptible to experimental test. Theorists have already suggested mass regions where it might be profitable to look for excited states of charmonium. Psi(3684) could be one of those states, but there would have to be more. The transition between psi(3684) and psi(3095) with the emission of two pions conforms to theoretical expectations. The energy transitions between other states have also been calculated, and if they exist, it should be possible to detect them. Finally, particles incorporating just one charmed quark

QUARK		CHARGE	BARYON NUMBER	CHARM	COLOR
u	u_r	$+2/3$	$+1/3$	0	RED
	u_y				YELLOW
	u_b				BLUE
d	d_r	$-1/3$	$+1/3$	0	RED
	d_y				YELLOW
	d_b				BLUE
s	s_r	$-1/3$	$+1/3$	0	RED
	s_y				YELLOW
	s_b				BLUE
c	c_r	$+2/3$	$+1/3$	$+1$	RED
	c_y				YELLOW
	c_b				BLUE

QUARK HYPOTHESIS states that hadrons are not elementary particles but composites of more fundamental entities called quarks. The original formulation of the theory, proposed independently by Murray Gell-Mann and George Zweig, postulated three quarks, *u*, *d* and *s*. Charge and baryon number (and other quantities not shown) are assigned to them according to the principle that baryons are made up of three quarks and mesons of a quark and an antiquark. Modifications of the theory add a fourth quark, *c*, which exhibits a property arbitrarily called charm, and propose that each quark exists in three states, distinguished by another property, called color. Thus there could be three, four, nine, 12 or more quarks.

MODEL OF HADRON STRUCTURE	PREDICTED VALUE OF HADRON-MUON PAIR RATIO
GELL-MANN-ZWEIG MODEL THREE QUARKS FRACTIONAL CHARGE	$\left(\frac{2}{3}\right)^2 + \left(-\frac{1}{3}\right)^2 + \left(-\frac{1}{3}\right)^2 = \frac{2}{3}$
COLOR HYPOTHESIS THREE TRIPLETS OF QUARKS FRACTIONAL CHARGE	$3\left[\left(\frac{2}{3}\right)^2 + \left(-\frac{1}{3}\right)^2 + \left(-\frac{1}{3}\right)^2\right] = 2$
COLOR HYPOTHESIS AND CHARM HYPOTHESIS THREE QUARTETS OF QUARKS FRACTIONAL CHARGE	$3\left[\left(\frac{2}{3}\right)^2 + \left(-\frac{1}{3}\right)^2 + \left(-\frac{1}{3}\right)^2 + \left(\frac{2}{3}\right)^2\right] = \frac{10}{3}$
HAN-NAMBU MODEL THREE QUARTETS OF QUARKS INTEGRAL CHARGE	$(1)^2 + (1)^2 + (-1)^2 + (-1)^2 = 4$

VARIANTS OF THE QUARK THEORY predict different values for the ratio of hadrons to muon pairs produced in electron-positron annihilations. In each case the predicted value is equal to the sum of the squares of the charges of the quarks included in the theory. The Gell-Mann–Zweig model gives a value of 2/3; the color hypothesis would increase the ratio to 2; assuming that both color and charm exist would yield a value of 10/3. The scheme devised by Moo-Young Han and Yoichiro Nambu eliminates the fractional charges common to other quark theories and predicts a value of 4 for the ratio. So far experimental findings at SPEAR and other laboratories cannot be reconciled with any of the predictions.

and therefore exhibiting charm should exist; the discovery of a charmed meson would provide strong evidence for the theory.

In experiments completed so far none of the phenomena that would confirm the color or charm hypotheses have been detected. Indeed, the failure to find this supporting evidence has become an embarrassment to both theories.

The new particles are not the only challenges to theory issuing from the recent discoveries. Some explanation is also required for the broad enhancement of hadron production in the vicinity of 4.1 GeV, and the questions raised by the odd behavior of the hadron/muon ratio throughout the energy range have not been resolved. The search for bumps and lines in the mass spectrum goes on not only at SPEAR but also at the DORIS storage ring at the German Electron Synchrotron in Hamburg and the ADONE storage ring at Frascati. Other investigators are studying the production of the psi particles through the interaction of photons and hadrons and through hadron-hadron collisions. (The last is the method employed by Ting and his colleagues at Brookhaven.)

In the experimental and theoretical investigations now under way many current concepts are being challenged; one, however, is not in question: that of quarks themselves. The discovery of the psi particles has confirmed again the central importance of quarks as the constituent particles of hadrons. Whether or not we shall ever see free quarks in the laboratory is another question; it is possible that they will always remain unobserved, exhibiting their physical reality only through their success in explaining the structure of hadrons and the forces that act on them.

Furthermore, we have no assurance that the quarks, whether there are three or nine or 12 or more of them, are the fundamental particles of matter. In the 20th century physics has probed the atom to discover the nucleus within, and has broken up the nucleus into its constituent particles. Those particles are now interpreted as being composites of more basic entities, the quarks. It is not unreasonable to imagine that we shall someday penetrate the quark and find an internal structure there as well. Only the experiments of the future can reveal whether quarks are the indivisible building blocks of all matter, the "atoms" of Democritus, or whether they too have a structure, as part of the endless series of seeds within seeds envisioned by Anaxagoras.

6

Quarks with Color and Flavor

by Sheldon Lee Glashow
October, 1975

*The particles called quarks may be truly elementary.
Their "colors" explain why they cannot be isolated;
their "flavors" distinguish four basic kinds, including
one that has the property called charm*

Atomos, the Greek root of "atom," means indivisible, and it was once thought that atoms were the ultimate, indivisible constituents of matter, that is, they were regarded as elementary particles. One of the principal achievements of physics in the 20th century has been the revelation that the atom is not indivisible or elementary at all but has a complex structure. In 1911 Ernest Rutherford showed that the atom consists of a small, dense nucleus surrounded by a cloud of electrons. It was subsequently revealed that the nucleus itself can be broken down into discrete particles, the protons and neutrons, and since then a great many related particles have been identified. During the past decade it has become apparent that those particles too are complex rather than elementary. They are now thought to be made up of the simpler things called quarks. A solitary quark has never been observed, in spite of many attempts to isolate one. Nonetheless, there are excellent grounds for believing they do exist. More important, quarks may be the last in the long series of progressively finer structures. They seem to be truly elementary.

When the quark hypothesis was first proposed more than 10 years ago, there were supposed to be three kinds of quark. The revised version of the theory I shall describe here requires 12 kinds. In the whimsical terminology that has evolved for the discussion of quarks they are said to come in four flavors, and each flavor is said to come in three colors. ("Flavor" and "color" are, of course, arbitrary labels; they have no relation to the usual meanings of those words.) One of the quark flavors is distinguished by the property called charm (another arbitrary term). The concept of charm was suggested in 1964, but until last year it had remained an untested conjecture.

Several recent experimental findings, including the discovery last fall of the particles called J or psi, can be interpreted as supporting the charm hypothesis.

The basic notion that some subatomic particles are made of quarks has gained widespread acceptance, even in the absence of direct observational evidence. The more elaborate theory incorporating color and charm remains much more speculative. The views presented here are my own, and they are far from being accepted dogma. On the other hand, a growing body of evidence argues that these novel concepts must play some part in the description of nature. They help to bring together many seemingly unrelated theoretical developments of the past 15 years to form an elegant picture of the structure of matter. Indeed, quarks are at once the most rewarding and the most mystifying creation of modern particle physics. They are remarkably successful in explaining the structure of subatomic particles, but we cannot yet understand why they should be so successful.

The particles thought to be made up of quarks form the class called the hadrons. They are the only particles that interact through the "strong" force. Included are the protons and neutrons, and indeed it is the strong force that binds protons and neutrons together to form atomic nuclei. The strong force is also responsible for the rapid decay of many hadrons.

Another class of particles, defined in distinction to the hadrons, are the leptons. There are just four of them: the electron and the electron neutrino and the muon and the muon neutrino (and their four antiparticles). The leptons are not subject to the strong force. Because the electron and the muon bear an electric charge, they "feel" the electromagnetic force, which is roughly 100 times weaker than the strong force. The two kinds of neutrino, which have no electric charge, feel neither the strong force nor the electromagnetic force, but interact solely through a third kind of force, weaker by several orders of magnitude and called the weak force. The strong force, the electromagnetic force and the weak force, together with gravitation, are believed to account for all interactions of matter.

The leptons give every indication of being elementary particles. The electron, for example, behaves as a point charge, and even when it is probed at the energies of the largest particle accelerators, no internal structure can be detected. The hadrons, on the other hand, seem complex. They have a measurable size: about 10^{-13} centimeter. Moreover, there are hundreds of them, all but a handful discovered in the past 25 years. Finally, all the hadrons, with the significant exception of the proton and the antiproton, are unstable in isolation. They decay into stable particles such as protons, electrons, neutrinos or photons. (The photon, which is the carrier of the electromagnetic force, is in a category apart; it is neither a lepton nor a hadron.)

The hadrons are subdivided into three families: baryons, antibaryons and mesons. The baryons include the proton and the neutron; the mesons include such particles as the pion. Baryons can be neither created nor destroyed except as pairs of baryons and antibaryons. This principle defines a conservation law, and it can be treated most conveniently in the system of bookkeeping that assigns simple numerical values, called quantum numbers, to conserved properties. In this case the quantum number is called baryon number. For

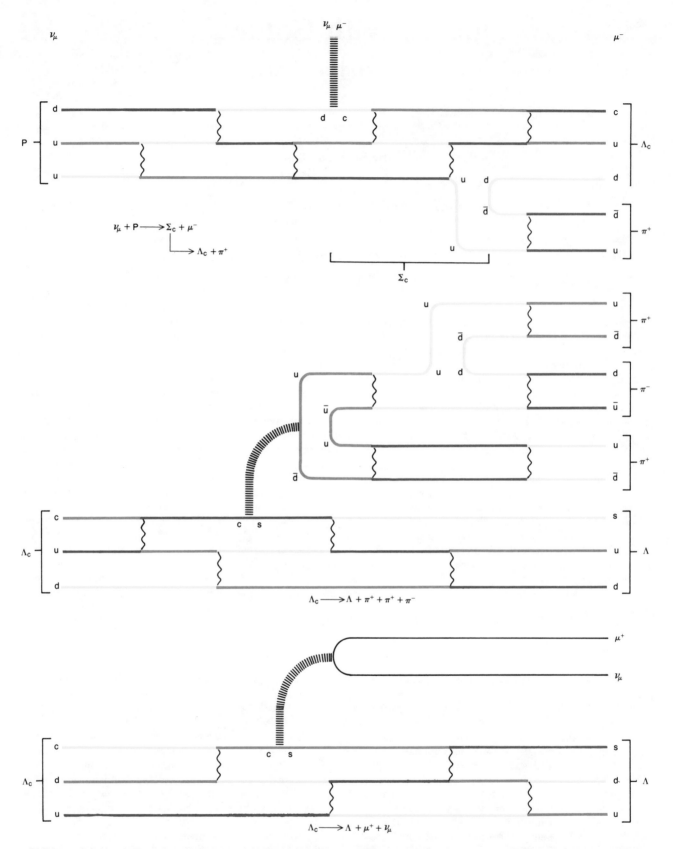

WEAK INTERACTIONS, mediated by the W particle (*hatched lines*), can change the flavor of a quark, but they have no effect on its color. Single charmed particles can therefore be created by the weak force, whereas they appear only as pairs in other interactions. A neutrino interacts with a proton (*top*), converting a d quark into a c quark while at the same time the neutrino is transformed into a muon. The immediate product is a charmed sigma baryon, but that quickly decays by the emission of a pion to a charmed lambda baryon. The charmed lambda particle can itself decay by emitting a W particle, converting the c quark into an s quark and thereby forming a strange lambda baryon. The strange lambda baryon can be accompanied by three pions, produced by repeated quark-pair creation (*middle*), a process that may have been observed recently at Brookhaven National Laboratory, or by a muon and a neutrino (*bottom*), a decay mode that may explain several events detected at the Fermi National Accelerator Laboratory.

baryons it is +1, for antibaryons −1 and for mesons 0. The conservation of baryon number then reduces to the rule that in any interaction the sum of the baryon numbers cannot change.

Baryon number provides a means of distinguishing baryons from mesons, but it is an artificial means, and it tells us nothing about the properties of the two kinds of particle. A more meaningful distinction can be established by examining another quantum number, spin angular momentum.

Under the rules of quantum mechanics a particle or a system of particles can assume only certain specified states of rotation, and hence can have only discrete values of angular momentum. The angular momentum is measured in units of $h/2\pi$, where h is Planck's constant, equal to about 6.6×10^{-27} erg-second. Baryons are particles with a spin angular momentum measured in half-integral units, that is, with values of half an odd integer, such as 1/2 or 3/2. Mesons have integral values of spin angular momentum, such as 0 or 1.

The difference in spin angular momentum has important consequences for the behavior of the two kinds of hadron. Particles with integral spin are said to obey Bose-Einstein statistics (and are therefore called bosons). Those with half-integral spin obey Fermi-Dirac statistics (and are called fermions). In this context "statistics" refers to the behavior of a population of identical particles. Those that obey Bose-Einstein statistics can be brought together without restriction; an unlimited number of pions, for example, can occupy the same state. The Fermi-Dirac statistics, on the other hand, require that no two particles within a given system have the same energy and be identical in all their quantum numbers. This statement is equivalent to the exclusion principle formulated in 1925 by Wolfgang Pauli. He applied it in particular to electrons, which have a spin of 1/2 and are therefore fermions. It requires that each energy level in an

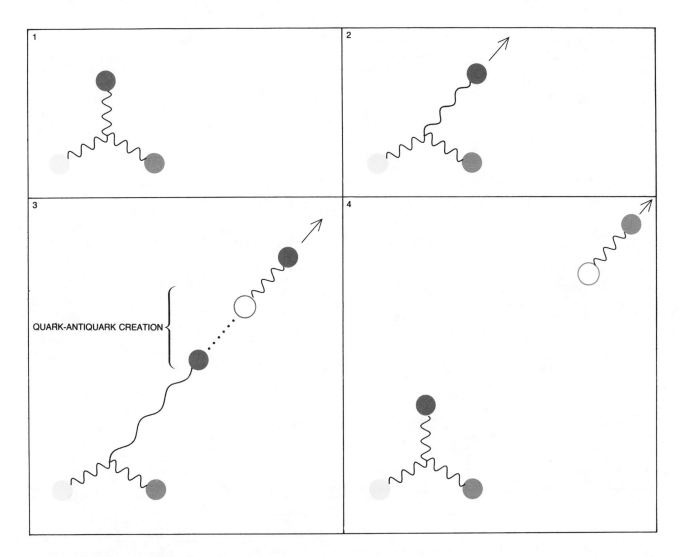

QUARK-ANTIQUARK CREATION

CONFINEMENT OF QUARKS in known subnuclear particles can be explained if the quarks are assumed to come in three varieties, arbitrarily designated by colors. A baryon (such as a proton or a neutron) is a bound system of three quarks, including one of each of the three colors (*1*), so that the baryon as a whole is colorless. The quarks are bound together by the strong force, which also holds together atomic nuclei. The force between colored quarks, however, differs in character from that between colorless composite particles: it does not diminish with distance but remains constant. As a result, when a quark is separated from a baryon (*2*), the potential energy of the system rapidly increases and would reach enormous values if another process did not intervene: from the potential energy a quark and an antiquark are created (*3*). The new quark restores the baryon to its original form, and the antiquark (*open circle*) adheres to the dislodged quark, forming another kind of particle, a meson (*4*). At any one moment the quark and the antiquark in a meson are the same color, but the three colors are equally represented. Because of the nature of the strong force, solitary quarks cannot be observed; any attempt to isolate a quark merely results in the creation of a new hadron or hadrons.

atom contain only two electrons, with their spins aligned in opposite directions.

One of the clues to the complex nature of the hadrons is that there are so many of them. Much of the endeavor to understand them has consisted of a search for some ordering principle that would make sense of the multitude.

The hadrons were first organized into small families of particles called charge multiplets or isotopic-spin multiplets; each multiplet consists of particles that have approximately the same mass and are identical in all their other properties except electric charge. The multiplets have one, two, three or four members. The proton and the neutron compose a multiplet of two (a doublet); both are considered to be manifestations of a single state of matter, the nucleon, with an average mass equivalent to an energy of .939 GeV (billion electron volts). The pion is a triplet with an average mass of .137 GeV and three charge states: +1, 0 and −1. In the strong interactions the members of a multiplet are all equivalent, since electric charge plays no role in the strong interactions.

In 1962 a grander order was revealed when the charge multiplets were organized into "supermultiplets" that revealed relations between particles that differ in other properties in addition to charge. The creation of the supermultiplets was proposed independently by Murray Gell-Mann of the California Institute of Technology and by Yuval Ne'eman of Tel-Aviv University [see the article "Strongly Interacting Particles," by Geoffrey F. Chew, Murray Gell-Mann and Arthur H. Rosenfeld, beginning on page 38.] The introduction of the new system led directly to the quark hypothesis.

The grouping of the hadrons into supermultiplets involves eight quantum numbers and has been referred to as the "eightfold way." Its mathematical basis is a branch of group theory invented in the 19th century by the Norwegian mathematician Sophus Lie. The Lie group that generates the eightfold way is called SU(3), which stands for the special unitary group of matrices of size 3×3. The theory requires that all hadrons belong to families corresponding to representations of the group SU(3). The families can have one, three, six, eight, 10 or more members. If the eightfold way were an exact theory, all the members of a given family would have the same mass. The eightfold way is only an approximation, however, and within the families there are significant differences in mass.

The construction of the eightfold way begins with the classification of the hadrons into broad families sharing a common value of spin angular momentum. Each family of particles with identical spin is then depicted by plotting the distribution of two more quantum numbers: isotopic spin and strangeness.

Isotopic spin has nothing to do with the spin of a particle; it was given its name because it shares certain algebraic properties with the spin quantum number. It is a measure of the number of particles in a multiplet, and it is calculated according to the formula that the number of particles in the multiplet is one more than twice the isotopic spin. Thus the nucleon (a doublet) has an isotopic spin of 1/2; for the pion triplet the isotopic spin is 1.

Strangeness is a quantum number introduced to describe certain hadrons first observed in the 1950's and called strange particles because of their anomalously long lifetimes. They generally decay in from 10^{-10} to 10^{-7} second. Although that is a brief interval by everyday standards, it is much longer than the lifetime of 10^{-23} second characteristic of many other hadrons.

Like isotopic spin, strangeness depends on the properties of the multiplet, but it measures the distribution of charge among the particles rather than their number. The strangeness quantum number is equal to twice the average charge (the sum of the charges divided by the number of particles in a multiplet) minus the baryon number. By this contrivance it is made to vanish for all hadrons except the strange ones. The triplet of pions, for example, has an average charge of 0 and a baryon number of 0; its strangeness is therefore also 0. The nucleon doublet has an average charge of + 1/2 and a baryon number of +1, so that those particles too have a strangeness of 0. On the other hand, the lambda particle is a neutral baryon that forms a family of one (a singlet). Its aver-

QUARKS					LEPTONS			
SYMBOL	MASS (GeV)	ELECTRIC CHARGE	STRANGE-NESS	CHARM	NAME	SYMBOL	MASS (GeV)	ELECTRIC CHARGE
d	.338	−1/3	0	0	ELECTRON	e^-	.0005	−1
u	.336	+2/3	0	0	ELECTRON NEUTRINO	ν_e	0	0
s	.540	−1/3	−1	0	MUON NEUTRINO	ν_μ	0	0
c	1.5	+2/3	0	+1	MUON	μ^-	.105	−1

QUARKS AND LEPTONS, the two kinds of particle that seem to be elementary, exhibit an apparent symmetry. The quarks are much more massive than the leptons, and they have fractional charges instead of integral ones, but both groups consist of two pairs of particles (indicated by colored rectangles). Either member of a pair is readily transformed into the other by the weak interaction. All ordinary matter can be constructed of just the d and u quarks and the electron and electron neutrino; the muon, the muon neutrino and the c and s quarks, which display the properties of strangeness and charm respectively, are important only in high-energy physics. Each kind, or flavor, of quark comes in three colors.

	NAME	SYMBOL	MASS (GeV)	CHARGE STATES	BARYON NUMBER	SPIN	ISOTOPIC SPIN	STRANGENESS
BARYONS	NUCLEON	N	.939	0, +1	+1	1/2	1/2	0
	LAMBDA	Λ	1.115	0	+1	1/2	0	−1
	OMEGA	Ω	1.672	−1	+1	3/2	0	−2
MESONS	PION	π	.139	−1, 0, +1	0	0	1	0
	K	K	.496	0, +1	0	0	1/2	+1
	PHI	φ	1.019	0	0	1	0	0
	J	J	3.095	0	0	1	0	0

HADRONS form the class of particles thought to be constructed of quarks. They are divided into baryons, made of three quarks, and mesons, made of a quark and an antiquark. (Antibaryons consist of three antiquarks.) The groups are distinguished by baryon number and by spin angular momentum, which has half-integral values for baryons and integral values for mesons. Each line in the table represents a multiplet of particles identical in all properties except electric charge, provided that small differences in mass are ignored. Isotopic spin is a function of the number of particles in a multiplet, and strangeness measures distribution of electric charge among them. Only a few representative hadrons are shown.

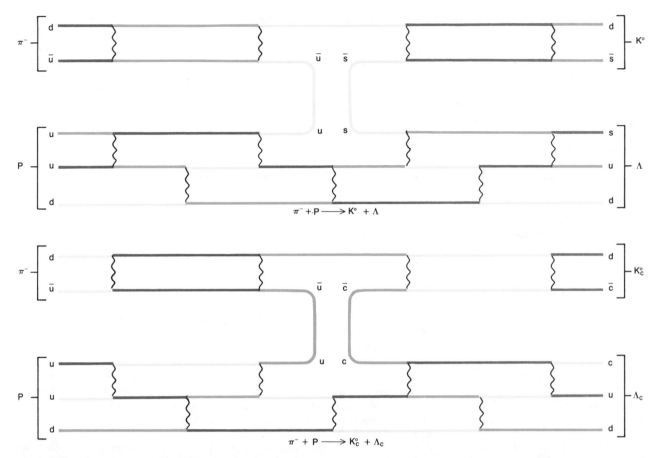

STRANGE AND CHARMED PARTICLES should be created in interactions of ordinary matter. The interactions are displayed here as intersections of lines representing quarks and other particles. Within a hadron the quarks continually exchange massless particles called gluons (*wavy lines*), the carriers of the strong force. By emitting a gluon a quark changes its color but not its flavor. Strange particles can be created (*top*) when a u quark in a proton and a $\bar{u}$ antiquark in a pion annihilate each other and give rise to an s quark and an $\bar{s}$ antiquark. Products are a K meson and a lambda baryon. At higher energy the same annihilation could yield a $\bar{c}$ quark and a $\bar{c}$ antiquark (*bottom*). This process, which has not been observed, would yield a charmed meson and a charmed baryon.

age charge of 0 and its baryon number of +1 give it a strangeness of −1.

On a graph that plots electric charge against strangeness the hadrons form orderly arrays. The mesons with a spin angular momentum of 0 compose an octet and a singlet; the octet is represented graphically as a hexagon with a particle at each vertex and two particles in the center, and the singlet is represented as a point at the origin. The mesons with a spin of 1 form an identical representation, and so do the baryons with a spin of 1/2. Finally, the baryons with a spin of 3/2 form a decimet (a group of 10) that can be graphed as a large triangle made up of a singlet, a doublet, a triplet and a quartet. The eightfold way was initially greeted with some skepticism, but the discovery in 1964 of the negatively charged omega particle, the predicted singlet in the baryon decimet, made converts of us all.

The regularity and economy of the supermultiplets are aesthetically satisfying, but they are also somewhat mystifying. The known hadrons do fit into such families, without exception. Mesons come only in families of one and eight, and baryons come only in families of one, eight and 10. The singlet, octet and decimet, however, are only a few of many possible representations of SU(3). Families of three particles or six particles are entirely plausible, but they are not observed. Indeed, the variety of possible families is in principle infinite. Why, then, do only three representations appear in nature? It early became apparent that the eightfold way is in some approximate sense true, but it was also plain from the start that there is more to the story.

In 1963 an explanation was proposed independently by Gell-Mann and by George Zweig, also of Cal Tech. They perceived that the unexpected regularities could be understood if all hadrons were constructed from more fundamental constituents, which Gell-Mann named quarks. The quarks were to belong to the simplest nontrivial family of the eightfold way: a family of three.

(There is also, of course, another family of three antiquarks.)

The quarks are required to have rather peculiar properties. Principal among these is their electric charge. All observed particles, without exception, bear integer multiples of the electron's charge; quarks, however, must have charges that are fractions of the electron's charge. Gell-Mann designated the three quarks u, d and s, for the arbitrary labels "up," "down" and "sideways."

The mechanics of the original quark model are completely specified by three simple rules. Mesons are invariably made of one quark and one antiquark. Baryons are invariably made of three quarks and antibaryons of three antiquarks. No other assemblage of quarks can exist as a hadron. The combinations of the three quarks under these rules are sufficient to account for all the hadrons that had been observed or predicted at the time. Furthermore, every allowed combination of quarks yields a known particle.

Many of the necessary properties of

the quarks can be deduced from these rules. It is mandatory, for example, that each of the quarks be assigned a baryon number of +1/3 and each of the antiquarks a baryon number of −1/3. In that way any aggregate of three quarks has a total baryon number of +1 and hence defines a baryon; three antiquarks yield a particle with a baryon number of −1, an antibaryon. For mesons the baryon numbers of the quarks (+1/3 and −1/3) cancel, so that the meson, as required, has a baryon number of 0.

In a similar way the angular momentum of the hadrons is described by giving the quarks half-integral units of spin. A particle made of an odd number of quarks, such as a baryon, must therefore also have half-integral spin, conforming to the known characteristics of baryons. A particle made of an even number of quarks, such as a meson, must have integral spin.

The *u* quark and the *s* quark compose an isotopic-spin doublet: they have nearly the same mass and they are identical in all other properties except electric charge. The *u* quark is assigned a charge of +2/3 and the *d* quark is assigned a charge of −1/3. The average charge of the doublet is therefore +1/6 and twice the average charge is +1/3; since the baryon number of all quarks is +1/3, the definition of strangeness gives both the *u* and the *d* quarks a strangeness of 0. The *s* quark has a larger mass than either the *u* or the *d* and makes up an isotopic-spin singlet. It is given an electric charge of −1/3 and consequently has a strangeness of −1. The antiquarks, denoted by writing the quark symbol with a bar over it, have opposite properties. The *ū* has a charge of −2/3 and the *d̄* +1/3; both have zero strangeness. The *s̄* antiquark has a charge of +1/3 and a strangeness of +1.

Just two of the quarks, the *u* and the *d*, suffice to explain the structure of all the hadrons encountered in ordinary matter. The proton, for example, can be described by assembling two *u* quarks and a *d* quark; its composition is written *uud*. A quick accounting will show that all the properties of the proton determined by its quark constitution are in accord with the measured values. Its charge is equal to 2/3 + 2/3 − 1/3, or +1. Similarly, its baryon number can be shown to be +1 and its spin 1/2. A positive pion is composed of a *u* quark and a *d̄* antiquark (written *ud̄*). Its charge is 2/3 + 1/3, or +1; its spin and baryon number are both 0.

The third quark, *s*, is needed only to construct strange particles, and indeed it provides an explicit definition of strangeness: A strange particle is one that contains at least one *s* quark or *s̄* antiquark. The lambda baryon, for example, can be shown from the charge distribution of its multiplet to have a strangeness of −1; that result is confirmed by its quark constitution of *uds*. Similarly, the neutral *K* meson, a strange particle, has a strangeness of +1, as confirmed by its composition of *ds̄*.

Until quite recently these three kinds of quark were sufficient to describe all the known hadrons. As we shall see, experiments conducted during the past year seem to have created hadrons whose properties cannot be explained in terms of the original three quarks. The experiments can be interpreted as implying the existence of a fourth kind of quark, called the charmed quark and designated *c*.

The statement that the *u*, *d* and *s* quarks are sufficient to construct all the observed hadrons can be made more precisely in the mathematical formalism of the eightfold way. Since a meson is made up of one quark and one antiquark, and since there are three kinds, or flavors, of quark, there are nine possible combinations of quarks and antiquarks that can form a meson. It can be shown that one of these combinations represents a singlet and the remaining eight form an octet. Similarly, since a baryon is made up of three quarks, there are 27 possible combinations of quarks that can make up a baryon. They can be broken up into a singlet, two octets and a decimet. Those groupings correspond exactly to the observed families of hadrons. The quark theory thus explains why only a few of the possible representations of SU(3) are realized in nature as hadron supermultiplets.

The quark rules provide a remarkably economical explanation for the formation of the observed hadron families. What principles, however, can explain the quark rules, which seem quite arbitrary? Why is it possible to bind together three quarks but not two or four? Why can we not create a single quark in isolation? A line of thought that leads to possible answers to these questions appeared at first as a defect in the quark theory.

As we have seen, it is necessary that the quarks have half-integral values of spin angular momentum; otherwise the known spins of the baryons and mesons would be predicted wrongly. Particles with half-integral spin are expected to obey Fermi-Dirac statistics and are therefore subject to the Pauli exclusion principle: No two particles within a particular system can have exactly the same quantum numbers. Quarks, however, seem to violate the principle. In making up a baryon it is often necessary that two identical quarks occupy the same state. The omega particle, for example, is made up of three *s* quarks, and all three must be in precisely the same state. That is possible only for particles that obey Bose-Einstein statistics. We are at an impasse: quarks must have half-integral spin but they must satisfy the statistics appropriate to particles having integral spin.

The connection between spin and statistics is an unshakable tenet of relativistic quantum mechanics. It can be deduced directly from the theory, and a violation has never been discovered. Since it holds for all other known particles, quarks could not reasonably be excluded from its dominion.

The concept that has proved essential to the solution of the quark statistics problem was proposed in 1964 by Oscar W. Greenberg of the University of Maryland. He suggested that each flavor of quark comes in three varieties, identical in mass, spin, electric charge and all other measurable quantities but different in an additional property, which has come to be known as color. The exclusion principle could then be satisfied, and quarks could remain fermions, because the quarks in a baryon would not all occupy the same state. The quarks could differ in color even if they were the same in all other respects.

The color hypothesis requires two additional quark rules. The first simply restates the condition that color was introduced to satisfy: Baryons must be made up of three quarks, all of which have different colors. The second describes the application of color to mesons: Mesons are made of a quark and an antiquark of the same color, but with equal representation of each of the three colors. The effect of these rules is that no hadron can exhibit net color. A baryon invariably contains quarks of each of the three colors, say red, yellow and blue. In the meson one can imagine the quark and antiquark as being a single color at any given moment, but continually and simultaneously changing color, so that over any measurable interval they will both spend equal amounts of time as red, blue and yellow quarks.

The price of the color hypothesis is a tripling of the number of quarks; there must be nine instead of three (with charm yet to be considered). At first it may also appear that we have greatly increased the number of hadrons, but that is an illusion. With color there seem to be nine times as many mesons and 27 times as many baryons, but the rules for assembling hadrons from colored quarks ensure that none of the additional particles are observable.

Although the quark rules imply that we will never see a colored particle, the color hypothesis is not merely a formal construct without predictive value. The increase it requires in the number of quarks can be detected in at least two ways. One is through the effect of color on the lifetime of the neutral pion, which almost always decays into two photons. Stephen L. Adler of the Institute for Advanced Study has shown that its rate of decay depends on the square of the number of quark colors. Just the observed lifetime is obtained by assuming that there are three colors.

Another effect of color can be detected in experiments in which electrons

and their antiparticles, the positrons, annihilate each other at high energy. The outcome of such an event is sometimes a group of hadrons and sometimes a muon and an antimuon. At sufficiently high energy the ratio of the number of hadrons to the number of muon-antimuon pairs is expected to approach a constant value, equal to the sum of the squares of the charges of the quarks. Tripling the number of quarks also triples the expected value of the ratio. The experimental result at energies of from 2 GeV to 3 GeV is in reasonable agreement with the color hypothesis (which predicts a value of 2) and is quite incompatible with the original theory of quarks without color.

The introduction of the color quantum number solves the problem of quark statistics, but it once again requires a set of rules that seem arbitrary. The rules

can be accounted for, however, by establishing another hypothetical symmetry group analogous to the SU(3) symmetry proposed by Gell-Mann and by Ne'eman. The earlier SU(3) is concerned entirely with combinations of the three quark flavors; the new one deals exclusively with the three quark colors. Moreover, unlike the earlier theory, which is only approximate, color SU(3) is supposed to be an exact symmetry, so that quarks of the same flavor but different color will have identical masses.

In the color SU(3) theory all the quark rules can be explained if we accept one postulate: All hadrons must be represented by color singlets; no larger multiplets can be allowed. A color singlet can be constructed in two ways: by combining an identically colored quark and antiquark with all three colors equally represented, or by combining three

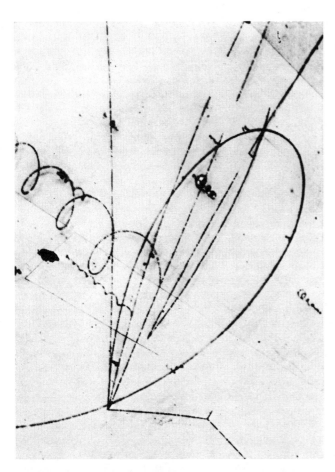

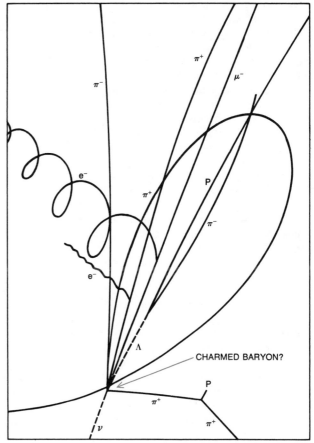

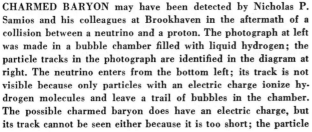

CHARMED BARYON may have been detected by Nicholas P. Samios and his colleagues at Brookhaven in the aftermath of a collision between a neutrino and a proton. The photograph at left was made in a bubble chamber filled with liquid hydrogen; the particle tracks in the photograph are identified in the diagram at right. The neutrino enters from the bottom left; its track is not visible because only particles with an electric charge ionize hydrogen molecules and leave a trail of bubbles in the chamber. The possible charmed baryon does have an electric charge, but its track cannot be seen either because it is too short; the particle

must decay in about 10^{-13} second, so that even at very high speed it does not move far enough to ionize more than a few molecules of hydrogen. The charmed particle decays into a neutral lambda particle, a strange baryon. The lambda particle leaves no track, but its decay products form a vertex that points toward the initial interaction. Four pions and a muon are also created, and two electrons struck by fast-moving particles spiral to the left in the bubble chamber's magnetic field. The presence of the charmed particle is not certain; several other interpretations of the event are possible, and although they are all unlikely, they cannot be excluded.

quarks or three antiquarks in such a way that the three colors are all included. These conditions, of course, are equivalent to the rules for building mesons, baryons and antibaryons, and they ensure that all hadrons will be colorless. There are no other ways to make a singlet in color SU(3); a particle made any other way would be a member of a larger multiplet, and it would display a particular color.

Although the color SU(3) theory of the hadrons can explain the quark rules, it cannot entirely eliminate the arbitrary element in their nature. We can ask a still more fundamental question: What explains the postulate that all hadrons must be color singlets? One approach to an answer, admittedly a speculative one, has been suggested recently by many investigators; it incorporates the color SU(3) model of the hadrons into one of the class of theories called gauge theories.

The color gauge theory postulates the existence of eight massless particles, sometimes called gluons, that are the carriers of the strong force, just as the photon is the carrier of the electromagnetic force. Like the photon, they are electrically neutral, and they have a spin of 1; they are therefore called vector bosons (bosons because they have integer spin and obey Bose-Einstein statistics, vector because a particle with a spin of 1 is described by a wave function that takes the form of a four-dimensional vector). Gluons, like quarks, have not been detected.

When a quark emits or absorbs a gluon, the quark changes its color but not its flavor. For example, the emission of a gluon might transform a red u quark into a blue or a yellow u quark, but it could not change it into a d or an s quark of any color. Since the color gluons are the quanta of the strong force, it follows that color is the aspect of quarks that is most important in the strong interactions. In fact, when describing interactions that involve only the strong force, one can virtually ignore the flavors of quarks.

The color gauge theory proposes that the force that binds together colored quarks represents the true character of the strong interaction. The more familiar strong interactions of hadrons (such as the binding of protons and neutrons in a nucleus) are manifestations of the same fundamental force, but the interactions of colorless hadrons are no more than a pale remnant of the underlying interaction between colored quarks. Just as the van der Waals force between molecules is only a feeble vestige of the electromagnetic force that binds electrons to nuclei, the strong force observed between hadrons is only a vestige of that operating within the individual hadron.

From these theoretical arguments one can derive an intriguing, if speculative, explanation of the confinement of quarks. It has been formulated by John Kogut and Kenneth Wilson of Cornell University and by Leonard Susskind of Yeshiva University. If it should be proved correct, it would show that the failure to observe colored particles (such as isolated quarks and gluons) is not the result of any experimental deficiency but is a direct consequence of the nature of the strong force.

The electromagnetic force between two charged particles is described by Coulomb's law: The force decreases as the square of the distance between the charges. Gravitation obeys a fundamentally similar law. At large distances both forces dwindle to insignificance. Kogut, Wilson and Susskind argue that the strong force between two colored quarks behaves quite differently: it does not diminish with distance but remains constant, independent of the separation of the quarks. If their argument is sound, an enormous amount of energy would be required to isolate a quark.

Separating an electron from the valence shell of an atom requires a few electron volts. Splitting an atomic nucleus requires a few million electron volts. In contrast to these values, the separation of a single quark by just an inch from the proton of which it is a constituent would require the investment of 10^{13} GeV, enough energy to separate the author from the earth by some 30 feet. Long before such an energy level could be attained another process would intervene. From the energy supplied in the effort to extract a single quark, a new quark and antiquark would materialize. The new quark would replace the one removed from the proton, and would reconstitute that particle. The new antiquark would adhere to the dislodged quark, making a meson. Instead of isolating a colored quark, all that is accomplished is the creation of a colorless meson [see illustration on page 73]. By this mechanism we are prohibited from ever seeing a solitary quark or a gluon or any combination of quarks or gluons that exhibits color.

If this interpretation of quark confinement is correct, it suggests an ingenious way to terminate the apparently infinite regression of finer structures in matter. Atoms can be analyzed into electrons and nuclei, nuclei into protons and neutrons, and protons and neutrons into quarks, but the theory of quark confinement suggests that the series stops there. It is difficult to imagine how a particle could have an internal structure if the particle cannot even be created.

Quarks of the same flavor but different color are expected to be identical in all properties except color; indeed, that is why the concept of color was introduced. Quarks that differ in flavor, however, have quite different properties. It is because the u quark and the d quark differ in electric charge that the proton is charged and the neutron is not. Similarly, it is because the s quark is considerably more massive than either the u or the d quark that strange particles are generally the heaviest members of their families. The charmed quark, c, must be heavier still, and charmed particles as a rule should therefore be heavier than all others. It is the flavor of quarks that brings variety to the world of hadrons, not their color.

As we have seen, the flavors of quarks are unaffected by the strong interactions. In a weak interaction, on the other hand, a quark can change its flavor (but not its color). The weak interactions also couple quarks to the leptons. The classical example of this coupling is nuclear beta decay, in which a neutron is converted into a proton with the emission of an electron and an antineutrino. In terms of quarks the reaction represents the conversion of a d quark to a u quark, accompanied by the emission of the two leptons.

The weak interactions are thought to be mediated by vector bosons, just as the strong and the electromagnetic interactions are. The principal one, labeled W and long called the intermediate vector boson, was predicted in 1938 by Hideki Yukawa. It has an electric charge of -1, and it differs from the photon and the color gluons in that it has mass, indeed a quite large mass. Quarks can change their flavor by emitting or absorbing a W particle. Beta decay, for example, is interpreted as the emission of a W by a d quark, which converts the quark into a u; the W then decays to yield the electron and antineutrino. From this process it follows that the W can also interact with leptons, and it thus provides a link between the two groups of apparently elementary particles.

The realization that the strong, weak and electromagnetic forces are all carried by the same kind of particle—bosons with a spin of 1—invites speculation that all three might have a common basis in some simple unified theory. A step toward such a unification would be the reconciliation of the weak interactions and electromagnetism. Julian Schwinger of Harvard University attempted such a unification in the mid-1950's (when I was one of his doctoral students, working on these very questions). His theory had serious flaws. One was eliminated in 1961, when I introduced a second, neutral vector boson, now called Z, to complement the electrically charged W. Other difficulties persisted for 10 years, until in 1967 Steven Weinberg of Harvard and Abdus Salam of the International Center for Theoretical Physics in Trieste independently suggested a resolution. By 1971 it was generally agreed, largely because of the work of Gerhard 't Hooft of the University of Utrecht, that the Weinberg-Salam conjecture is successful [see the article "Unified Theories of Elementary-Particle Interaction," by Steven Weinberg, beginning on page 110].

Through the unified weak and electromagnetic interactions, quarks and leptons are intimately related. These interactions "see" the four leptons and distinguish between the three quark flavors. The W particle can induce one kind of neutrino to become an electron and the other kind of neutrino to become a muon. Similarly, the W can convert a u quark into a d quark; it can also influence the u quark to become an s quark, although much less readily.

There is an obvious lack of symmetry in these relations. The leptons consist of two couples, married to each other by the weak interaction: the electron with the electron neutrino and the muon with the muon neutrino. The quarks, on the other hand, come in only three flavors, and so one must remain unwed. The scheme could be made much tidier if there were a fourth quark flavor, in order to provide a partner for the unwed quark. Both the quarks and the leptons would then consist of two pairs of particles, and each member of a pair could change into the other member of the same pair simply by emitting a W. The desirability of such lepton-quark symmetry led James Bjorken and me, among others, to postulate the existence of a fourth quark in 1964. Bjorken and I called it the charmed quark. When provisions are made for quark colors, charm becomes a fourth quark flavor, and a new triplet of colored quarks is required. There are thus a total of 12 quarks.

Since 1964 several additional arguments for charm have developed. To me the most compelling of them is the need to explain the suppression of certain in-

teractions called strangeness-changing neutral currents. An explanation that relies on the properties of the charmed quark was presented in 1967 by John Iliopoulos, Luciano Maiani and me.

Strangeness-changing neutral currents are weak interactions in which the net electric charge of the hadrons does not change but the strangeness does; typically an s quark is transformed into a d quark, and two leptons are emitted. An example is the decay of the neutral K meson (a strange particle) into two oppositely charged muons. Such processes are found by experiment to be extremely rare. The three-quark theory cannot account for their suppression, and in fact the unified theory of weak and electromagnetic interactions predicts rates more than a million times greater than those observed.

The addition of a fourth quark flavor with the same electric charge as the u quark neatly accounts for the suppression, although the mechanism by which it does so may seem bizarre. With two pairs of quarks there are two possible paths for the strangeness-changing interactions, instead of just one when there are only three quarks. In the macroscopic world the addition of a second path, or channel, would be expected always to bring an increase in the reaction rate. In a world governed by quantum mechanics, however, it is possible to subtract as well as to add. As it happens, a sign in the equation that defines one of the reactions is negative, and the two interactions cancel each other.

The addition of a fourth quark flavor must obviously increase the number of hadrons. In order to accommodate the newly predicted particles in supermultiplets the eightfold way must be expanded. In particular another dimension must be added to the graphs employed to represent the families, so that the plane figures of the earlier symmetry become Platonic and Archimedean solids.

To the meson octet are added six charmed particles and one uncharmed particle to make up a new family of 15. It is represented as a cuboctahedron, in which one plane contains the hexagon of the original uncharmed meson octet. The baryon octets and decimet are expected to form two families having 20 members each. They are represented as a tetrahedron truncated at each vertex and as a regular tetrahedron. In addition there is a smaller regular tetrahedron consisting of just four baryons. Again, each figure contains one plane of uncharmed particles [see illustration on page 83].

It now appears that the first of the

new particles to be discovered is a meson that is not itself charmed. That conclusion is based on the assumption that the predicted meson is the same particle as the J or psi particle discovered last November. The announcement of the discovery was made simultaneously by Samuel C. C. Ting and his colleagues at the Brookhaven National Laboratory and by Burton Richter, Jr., and a group of other physicists at the Stanford Linear Accelerator Center (SLAC). At Brookhaven it was named J, at Stanford psi. Here I shall adopt the name J. For two excited states of the same particle, however, the names psi' and psi" will be employed, since they were seen only in the SLAC experiments.

The J particle was found as a resonance, an enhancement at a particular energy in the probability of an interaction between other particles. At Brookhaven the resonance was detected in the number of electron-positron pairs produced in collisions between protons and atomic nuclei. At SLAC it was observed in the products of annihilations of electrons and positrons. The energy at which the resonances were observed—and thus the energy or mass of the J particle—is about 3.1 GeV [see the article "Electron-Positron Annihilation and the New Particles," by Sidney D. Drell, beginning on page 58].

The J particle decays in about 10^{-20} second, certainly a brief interval, but nevertheless 1,000 times longer than the expected lifetime of a particle having the J's mass. The considerable excitement generated by the discovery of the J was largely a result of its long lifetime.

A great many explanations of the particle were proposed; for example, it was suggested that it might be the Z. I believe there is good reason to interpret the J as being a meson made up of a charmed quark and a charmed antiquark, that is, a meson with the quark constitution $c\bar{c}$. Thomas Appelquist and H. David Politzer of Harvard have named such a meson "charmonium," by analogy to positronium, a bound state of an electron and a positron. Charmonium is without charm because the charm quantum numbers of its quarks (+1 and −1) add up to zero.

The charmonium hypothesis can account for the anomalous lifetime of the J if one considers the ultimate fate of the decaying particle's quarks. There are three possibilities: they can be split up to become constituents of two daughter hadrons, they can both become part of a single daughter particle or they can be annihilated. An empirical rule, first

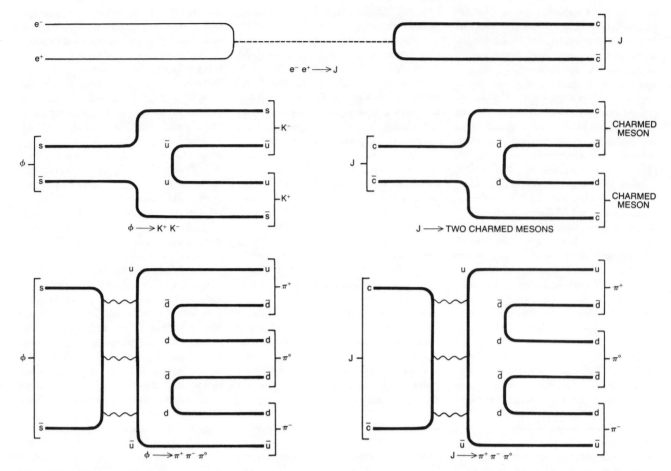

J PARTICLE is interpreted as a bound state of a c quark and a c̄ antiquark, called charmonium. It was discovered last November by physicists at Brookhaven and at the Stanford Linear Accelerator Center (SLAC). At SLAC it was made (*top*) when electrons and positrons annihilated each other to create a virtual photon (*broken line*), which then materialized to yield the new meson. The J particle has a mass of 3.1 GeV (billion electron volts). It does not exhibit charm because the charm quantum numbers of its quarks (+1 and −1) add to zero; in the same way a more familiar particle, the phi meson, has zero strangeness because it is made up of a strange quark and a strange antiquark. In spite of this analogy the two particles decay by different modes. For both the preferred mode consists in contributing the quark and the antiquark to two different daughter mesons. For the phi meson that is just possible (*middle left*) because the mass of the phi (1.019 GeV) is slightly greater than the combined mass of two strange K mesons. For the J it is not possible (*middle right*) because the lightest charmed particle is more than half as massive as the J. The J must therefore decay by the annihilation of its quarks (*bottom right*); the annihilation yields three gluons (*wavy lines*), which are transformed into three pions. This mode of decay is suppressed, and the equivalent process for the phi meson (*bottom left*) is rarely observed.

noted by Zweig, states that decays of the first kind are allowed but the other two are suppressed. For the J particle to decay in the allowed manner it must create two charmed particles, that is, two hadrons, one containing a charmed quark and the other a charmed antiquark. That decay is possible only if the mass of the J is greater than the combined masses of the charmed daughter particles. There is reason to believe the lightest charmed particle has a mass greater than half of the mass of the J, and it therefore appears that the J cannot decay in the allowed mode. The J cannot decay in the second way, either, keeping both its quarks in a single particle, because the J is the least massive state containing a charmed quark and a charmed antiquark. It must therefore decay by the annihilation of its quarks, a decay suppressed by Zweig's rule. The suppression

offers a partial explanation for the particle's extended lifetime.

Zweig's rule was formulated to explain the decay of the phi meson, which is made up of a strange quark and a strange antiquark and has a mass of about 1 GeV. The two particles are closely analogous, but the decay of the J is appreciably slower than that of the phi. Why should Zweig's rule be more effective for J than it is for phi? Furthermore, what explains Zweig's arbitrary rule?

A possible answer is provided by the theoretical concept called asymptotic freedom, which holds that the strong interactions become less strong at high energy. At sufficiently high energy the proton behaves as if it were made up of three freely moving quarks instead of three tightly bound ones. The concept takes its name from the fact that the

quarks approach the state of free motion asymptotically as the energy is increased. Asymptotic freedom offers an explanation for the discrepancy between the phi and the J particles in the application of Zweig's rule. Because the J is so massive, or alternatively so energetic, the strong interaction is of diminished strength, and it is particularly difficult for the quark and the antiquark to annihilate each other.

Like positronium, charmonium should appear in many energy states. Two were discovered at SLAC soon after the first state was found; they are psi′, with a mass of about 3.7 GeV, and psi″, with a mass of about 4.1 GeV. They appear to be simple excited states of the lowest-lying state of charmonium, the J particle. Psi′ decays only a little more quickly than J, and half the time its decay products are the J particle itself and two pions. Thus it sometimes decays by the second suppressed process described by Zweig's rule, that is, by contributing both of its quarks to a single daughter particle. The extended lifetime implies that psi' also lies below the energy threshold for the creation of a pair of charmed particles.

Psi'' decays much more quickly and therefore must be decaying in some mode permitted by Zweig's rule. Its decay products have not yet been determined, but it is possible they include charmed hadrons.

Numerous other excited states of charmonium follow inevitably from the theory of quark interactions [*see illustration at right*]. One, called *p*-wave charmonium, is formed when the particle takes on an additional unit of angular momentum. Some fraction of the time psi' should decay into *p*-wave charmonium, which should subsequently decay predominantly to the ground state, *J*. At each transition a photon of characteristic energy must be emitted. Recent experiments at the DORIS particle-storage rings of the German Electron Synchrotron in Hamburg have apparently detected the decays associated with the *p*-wave particle. In a few percent of its decays psi' yields the *J* particle and two photons, with energies of .2 GeV and .4 GeV. At SLAC psi' has been found to decay into an intermediate state and a single photon with an energy of .2 GeV. The intermediate state, which is presumably the same particle as the one observed at DORIS, then decays directly into hadrons.

The correspondence of theory and experiment revealed by the discovery of the *p*-wave transitions inspires considerable confidence that the charmonium interpretation of the *J* particle is correct. There is at least one more predicted state, called paracharmonium, that must be found if this explanation of the particle is to be confirmed. It differs from the observed states in the orientation of the quark spins: in *J*, psi' and psi'' (collectively called orthocharmonium) they are parallel; in paracharmonium they are antiparallel. Paracharmonium has so far evaded detection, but if the theoretical description is to make sense, paracharmonium must exist.

In addition to the various states of (uncharmed) charmonium, all the predicted charmed particles must also exist. If the *J* is in fact a state of charmonium, we can deduce from its mass the masses of all the hadrons containing charmed quarks.

An important initial constraint on the

range of possible masses was provided by the interpretation of the suppression of strangeness-changing neutral currents. If the suppression mechanism is to work, the charmed quark cannot be too much heavier than its siblings. On the other hand, it cannot be very light or charmed hadrons would already have been observed. An estimate from these conditions suggested that charmed particles would be found to have masses of about 2 or 3 GeV.

After the discovery of the *J*, I performed a more formal analysis with my colleagues at Harvard, Alvaro De Rújula and Howard Georgi. So did many others. Our estimates indicate that the least massive charmed states are mesons made up of a *c* quark and a *ū* or *d̄* antiquark; their mass should fall between 1.8 GeV and 2.0 GeV. A value within that range could be in agreement with the supposition that psi' lies below the threshold for the creation of a pair of charmed mesons, but psi'' lies above it.

The least massive charmed baryon has a quark composition of *udc*; we predict that its mass is near 2.2 GeV. As might

be expected, since the *c* quark is the heaviest of the four, the most massive predicted charmed hadron is the *ccc* baryon. We estimate its mass at about 5 GeV.

An important principle guiding experimental searches for charmed hadrons is the requirement that in most kinds of interaction charmed particles can be created only in pairs. Two hadrons must be produced, one containing a charmed quark, the other a charmed antiquark; the obvious consequence is a doubling of the energy required to create a charmed particle. An important exception to this rule is the interaction of neutrinos with other kinds of particles, such as protons. Neutrino events are exempt because neutrinos have only weak interactions and quark flavor can be changed in weak processes. Many experimental techniques have been tried in the search for charm during the past 10 years, yet no charmed particle has been unambiguously identified. Nevertheless, two recent experiments, both involving neutrino interactions, are encouraging. In both charm may at last have appeared,

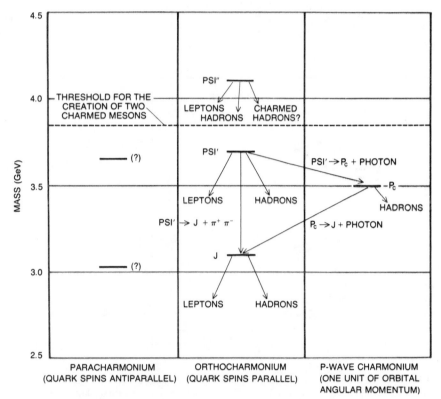

CHARMONIUM must exist at several energy levels, distinguished by the state of motion of the constituent quarks. The **J** particle is the ground state of orthocharmonium, in which the quark spins are parallel. Two excited states of orthocharmonium, designated psi' and psi'', were discovered at SLAC shortly after the **J** particle. Psi', like the **J** particle, seems to be too light to decay into two charmed hadrons, but the rapid decay of psi'' suggests that it has the necessary mass. Two other forms of the particle, called paracharmonium, in which the quark spins are antiparallel, have not been discovered. *P*-wave charmonium, in which the quarks have a unit of orbital angular momentum in addition to spin angular momentum, may have been detected at the German Electron Synchrotron in Hamburg and at SLAC.

but even if that proves to be an illusion, the experiments suggest promising lines of research.

One of the experiments was conducted at the Fermi National Accelerator Laboratory in Batavia, Ill., by a group of physicists headed by David B. Cline of the University of Wisconsin, Alfred K. Mann of the University of Pennsylvania and Carlo Rubbia of Harvard. In examining the interactions of high-energy neutrinos they found that in several percent of the events the products included two oppositely charged muons. One of the muons could be created directly from the incident neutrino, but the other is difficult to account for with only the ensemble of known, uncharmed particles. The most likely interpretation is that a heavy particle created in the reac-

tion decays by the weak force to emit the muon. The particle would have a mass of between 2 and 4 GeV, and if it is a hadron, some explanation must be found for its weak decay. Most particles with masses that large decay by the strong force. The presence of a charmed quark in the particle might provide the required explanation.

The second experiment was performed at Brookhaven by a group of investigators under Nicholas P. Samios. They photographed the tracks resulting from the interaction of neutrinos with protons in a bubble chamber. In a sample of several hundred observed collisions one photograph seemed to have no conventional interpretation [see illustration on page 77]. The final state can be construed as the decay products of a

charmed baryon. The process would provide convincing evidence for the existence of charm if it were not attested to by only one event. A few more observations of the same reaction would settle the matter.

It would be misleading to give the impression that the description of hadrons in terms of quarks of three colors and four flavors has solved all the outstanding problems in the physics of elementary particles. For example, continuing measurements of the ratio of hadrons to muon pairs produced in electron-positron annihilations have confounded prediction. The ratio discriminates between various quark models, and an argument in support of the color hypothesis was that at energies of from 2 to 3 GeV the ratio is about 2. At higher energy, high enough for charmed hadrons to be created in pairs, the ratio was expected to rise from 2 to about 3.3. The ratio does increase, but it overshoots the mark and appears to stabilize at a value of about 5. Perhaps charmed particles are being formed, but it seems that something else is happening as well: some particle is being made that does not appear in the theory I have described. One of my colleagues at Harvard, Michael Barnett, believes we have not been ambitious enough. He invokes six quark flavors rather than four, so that there are three flavors of charmed quark. It is also possible there are heavier leptons we know nothing about.

Finally, even if a completely consistent and verifiable quark model could be devised, many fundamental questions would remain. One such perplexity is implicit in the quark-lepton symmetry that led to the charm hypothesis. Both the quarks and the leptons, all of them apparently elementary, can be divided into two subgroups. In one group are the u and d quarks and the electron and electron neutrino. These four particles are the only ones needed to construct the world; they are sufficient to build all atoms and molecules, and even to keep the sun and other stars shining. The other subgroup consists of the strange and charmed quarks and the muon and muon neutrino. Some of them are seen occasionally in cosmic rays, but mainly they are made in high-energy particle accelerators. It would appear that nature could have made do with half as many fundamental things. Surely the second group was not created simply for the entertainment or edification of physicists, but what is the purpose of this grand doubling? At this point we have no answer.

CHARMED BARYONS are expected to be considerably more massive than other hadrons. None of the charmed particles have yet been unambiguously identified, but their masses have been predicted from the mass of the J particle. Some of the charmed particles must exist in more than one charge state (indicated by zeros and plus signs) and at several energy levels (indicated by their position with respect to the mass scale at left). Some of the particles can decay by the strong interaction (*dotted arrows*) or the electromagnetic interaction (*solid black arrows*) into states that have the same quantum numbers but smaller mass; others can decay only by the weak interaction (*open arrows*) into uncharmed particles. The form of the table is determined largely by the requirement that a baryon be made up of exactly three quarks; there can be no particle with a strangeness of −2 and a charm of +3, for example, because that would require five quarks: two strange ones and three charmed ones.

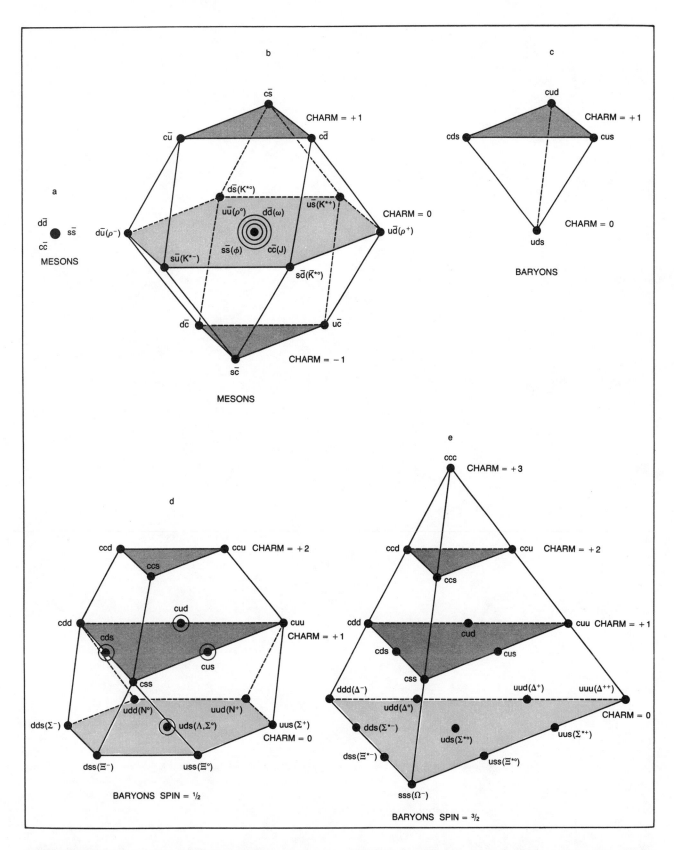

SUPERMULTIPLETS OF HADRONS that include the predicted charmed particles can be arranged as polyhedrons. Each supermultiplet consists of particles with the same value of spin angular momentum. Within each supermultiplet the particles are assigned positions according to three quantum numbers: positions on the shaded planes are determined by isotopic spin and strangeness; the planes themselves indicate values of charm. The mesons are represented by a point (*a*) and by an Archimedean solid called a cuboctahedron (*b*), which comprises 15 particles, including six charmed ones. The mesons shown are those with a spin of 1, but all mesons fit the same point and cuboctahedron representations. The baryons form a small regular tetrahedron (*c*) of four particles, a truncated tetrahedron (*d*) of 20 particles and a larger regular tetrahedron (*e*), also made up of 20 particles. Both mesons and baryons are identified by their quark constitution, and for those particles that have been observed the established symbol is also given. Each figure contains one plane (*color*) of uncharmed particles that are identical with earlier representations of the "eightfold way."

7

The Confinement of Quarks

by Yoichiro Nambu
November, 1976

How is it that these elementary particles of matter that explain so much about other particles are not seen? It may be that they are held inside other particles by forces inherent in their nature

An elementary particle of matter, strictly defined, is one that has no internal structure, one that cannot be broken up into smaller constituent particles. In the past decade or so it has become apparent that many particles long thought to be elementary, including such familiar ones as the proton and the neutron, are not elementary at all. Instead they appear to be composite structures made up of the more fundamental entities named quarks, in much the same way that an atom is made up of a nucleus and electrons.

The quark model amounts to an impressive simplification of nature. In the initial formulation of the theory there were supposed to be just three species of quark, and those three were enough to account for the properties of an entire class of particles with several dozen members. Every known member of that class could be understood as a combination of quarks; moreover, every allowed combination of the quarks gave rise to a known particle. The correspondence between theory and observation seemed too close to be coincidental, and experiments were undertaken with the aim of detecting the quarks themselves.

If the quarks are real particles, it seems reasonable that we should be able to see them. We know that the atom consists of a nucleus and a surrounding cloud of electrons because we can take the atom apart and study its constituents in isolation. We know that the nucleus in turn consists of protons and neutrons because the nucleus can be split into fragments and the constituent particles identified. It is easy to imagine a similar experiment in which particles thought to be made of quarks, such as protons, are violently decomposed. When that is attempted, however, the debris consists only of more protons and other familiar particles. No objects with the properties attributed to quarks are seen. Physicists have searched high and low, but free quarks have not been found.

It is possible, of course, that no experiment has yet looked in the right place or with the right instruments, but that now seems unlikely. It is also possible that the quarks simply do not exist, but physicists are reluctant to abandon a theory that carries such explanatory force. The successes of the theory represent compelling evidence that quarks exist inside particles such as the proton; on the other hand, the repeated failure of experimental searches to discover a free quark argues that the quarks do not exist independently. This paradox can be resolved, but only by making further theoretical assumptions about the quarks and the forces that bind them together. It must be demonstrated that quarks exist, but that for some reason they do not show themselves in the open. Theorists, who invented the quarks in the first place, are now charged with explaining their confinement within the particles they make up.

The particles thought to be made up of quarks are those called hadrons; they are distinguished by the fact that they interact with each other through the strong force, the force that binds together the particles in the atomic nucleus. ("Hadron" is derived from the Greek *hadros,* meaning stout or strong.) No other particles respond to the strong force.

The hadrons are divided into the two large subgroups named the baryons and the mesons. These two kinds of particle differ in many of their properties, and indeed they play different roles in the structure of matter, but the distinction between them can be made most clearly in the context of the quark model. All baryons consist of three quarks, and there are also antibaryons consisting of three antiquarks. The proton and the neutron are the least massive and the most familiar of the baryons. Mesons have a different structure: they consist of a quark bound to an antiquark. The pi meson, or pion, is the least massive of the mesons.

The properties of the hadrons can perhaps be best illustrated by considering them in contrast with the remaining major group of particles: the leptons. The leptons are not susceptible to the strong force (or else they would be hadrons). There are just four of them: the electron, the muon, the electron-type neutrino and the muon-type neutrino (along with the four corresponding antiparticles). The leptons seem to be truly elementary: they have no internal structure. Indeed, they apparently have no size. They can be represented as dimensionless points, so that it would not seem possible for them to have an internal structure.

The hadrons differ from the leptons in many respects, and they give several clues to their composite nature. The hadrons have a finite size, even if it is exceedingly small: about 10^{-13} centimeter. Experiments in which protons or neutrons collide at high energy with other particles give fairly direct evidence of an internal structure: electric and magnetic fields and the field associated with the strong force all seem to emanate from point sources within the particles. Finally, there are a great many hadrons. Well over 100 are known, most of them very short-lived, and there is every indication that many more exist and have not yet been observed only because the particle accelerators available today cannot supply enough energy to create them. There is no obvious limit to the number of hadrons that might be found as larger accelerators are built.

It was the great multiplicity of the hadrons that led to the formulation of the quark model. Without some organizing principle such a large collection of particles seemed unwieldy, and the possibility that they might all be elementary offended those who hold the conviction, or at least the fond wish, that nature should be simple. The quark hypothesis replaced the great variety of hadrons with just three fundamental building blocks from which all the hadrons could be constructed. It was proposed in 1963, independently by Murray Gell-Mann and George Zweig, both of the California Institute of Technology. It was Gell-Mann who supplied the name quark, from a line in James Joyce's *Finnegans Wake:* "Three quarks for Muster Mark!"

The immediate inspiration for the quark hypothesis was the discovery, made by Gell-Mann and by Yuval Ne'eman of Tel-Aviv University, that all the hadrons can be grouped logically in families of a few members each. The mesons form families of one particle and of eight particles; the baryons form families of one, eight and 10.

The classification of particles is made easier by tabulating their properties in numerical form. Each number refers to a single property, and it can assume only certain discrete values. Since the numbers are assigned in discrete units, or quanta, they are called quantum numbers. A complete list of a particle's quantum numbers identifies it uniquely and defines its behavior.

Electric charge is a typical quantum number. The fundamental unit of measurement is the electric charge of the proton or the electron, and in those units the charges of all observed particles can be expressed as simple integers (such as 0, $+1$ and -1). Another quantum number is called baryon number. Baryons are arbitrarily assigned a value of $+1$ and antibaryons a value of -1. Mesons have a baryon number of 0. Strangeness, the property of hadrons introduced in the 1950's to explain the strangely long lifetimes of some massive particles, is also accounted for through a quantum number with only integer quantities.

One of the most important quantum numbers is spin angular momentum. Under the rules of quantum mechanics a particle's state of rotation is one of its intrinsic properties, and the particle must therefore always have a specified and unvarying angular mo-

mentum. (The angular momentum is measured in units of Planck's constant divided by 2π, Planck's constant being 6.6×10^{-27} erg-second.) A crucial distinction is made between particles whose spins have a half-integer value (that is, half an odd integer, such as $1/2$ or $3/2$) and those with integer spin (such as 0, 1 or 2). As we shall see, this distinction determines the behavior of the particles when they are brought together in a bound system, but for now it is enough to note that all baryons have half-integer spins and all mesons have integer spins.

The families of hadrons defined by Gell-Mann and Ne'eman are related by spin angular momentum. All the members of a given family have the same quantity of spin. Within the families the members are distinguished from one another by two other quantum numbers: isotopic spin and hypercharge. In spite of its name, isotopic spin has nothing to

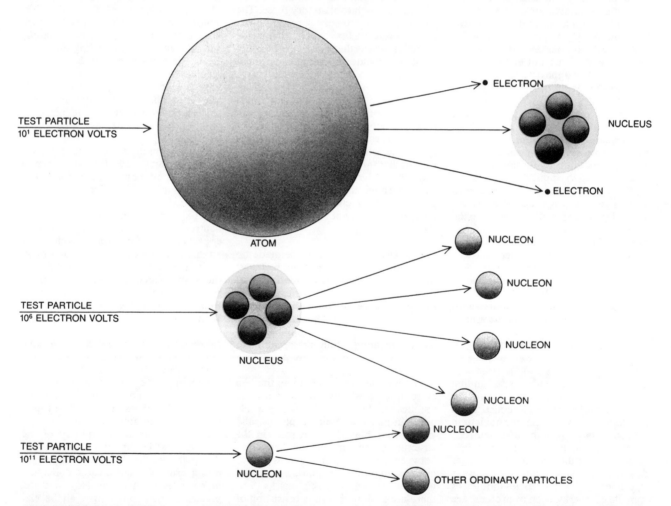

TEST PARTICLE
10^1 ELECTRON VOLTS

ATOM

ELECTRON

NUCLEUS

ELECTRON

TEST PARTICLE
10^6 ELECTRON VOLTS

NUCLEUS

NUCLEON

NUCLEON

NUCLEON

NUCLEON

TEST PARTICLE
10^{11} ELECTRON VOLTS

NUCLEON

NUCLEON

OTHER ORDINARY PARTICLES

STRUCTURE OF MATTER has been examined at progressively finer scale by a process of violent decomposition. The atom can be reduced to its component parts by striking it with a projectile carrying relatively little energy: a few electron volts. This is the process called ionization, and in the extreme case it results in the isolation of free electrons and a nucleus. The nucleus can also be dismembered, although higher energy is required. The nucleus splits into free protons and neutrons, which are collectively called nucleons. Nucleons in turn seem to be composed of the pointlike entities called quarks, and it would seem that the quarks might be liberated by smashing a nucleon with a test particle of sufficient energy. When that experiment is attempted, however, free quarks are not seen, even at the highest energies now attainable (a few hundred billion electron volts). Instead other ordinary particles are created, including many that are thought to be made up of quarks. A possible explanation of this effect is that the quarks are permanently confined within the nucleon.

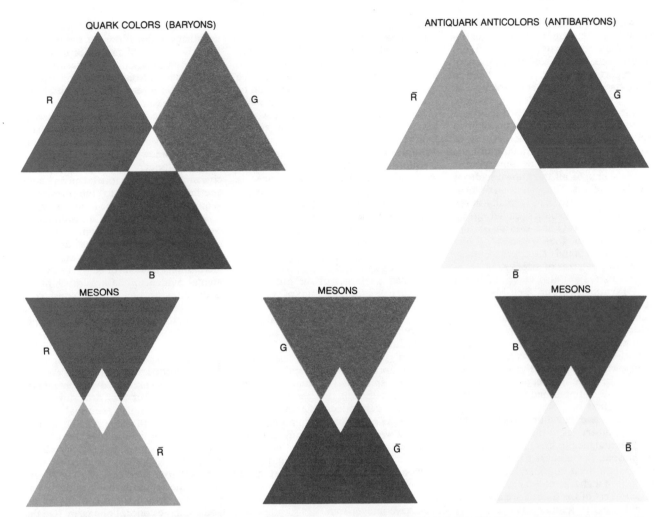

QUARK COLORS (BARYONS)

R G

B

ANTIQUARK ANTICOLORS (ANTIBARYONS)

R̄ Ḡ

B̄

MESONS

R

R̄

MESONS

G

Ḡ

MESONS

B

B̄

COLORLESS HADRONS are formed by properly combining the colored quarks. A baryon consists of three quarks, one each of red, green and blue color. (The quarks can have any flavor, and their flavors determine all the observable properties of the particle.) Similarly, an antibaryon is made up of three antiquarks having each of the three anticolors. The anticolors are shown here as the complements of the corresponding primary colors. In mesons the colors and anticolors are equally represented. In each of these combinations the net color quantum numbers are zero; in figurative terms the hadrons are white or colorless. No other combinations of colors can give the same result. It is thus possible to explain why only these combinations of quarks exist in nature, and why single quarks are prohibited, by postulating that only colorless particles are observable. The problem of quark confinement is then reduced to the problem of explaining this postulate.

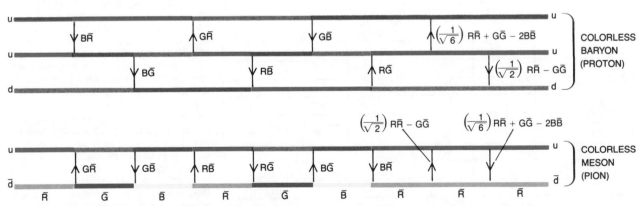

$\left(\frac{1}{\sqrt{6}}\right) R\bar{R} + G\bar{G} - 2B\bar{B}$ — COLORLESS BARYON (PROTON)

$\left(\frac{1}{\sqrt{2}}\right) R\bar{R} - G\bar{G}$

$\left(\frac{1}{\sqrt{2}}\right) R\bar{R} - G\bar{G}$ $\left(\frac{1}{\sqrt{6}}\right) R\bar{R} + G\bar{G} - 2B\bar{B}$ — COLORLESS MESON (PION)

EXCHANGE OF GLUONS binds together the quarks in a hadron and can simultaneously change their colors. In these diagrams the vertical dimension represents the spatial separation between the quarks and the horizontal dimension represents time. At each vertex where a gluon is emitted or absorbed the color quantum numbers must balance. Hence at the upper left, where a blue quark emits a blue-antired gluon, the "bluishness" of the quark is carried off by the gluon, and the quark changes to red, balancing the antired of the gluon. When the gluon is absorbed, the antired of the gluon and the red of the absorbing quark annihilate each other, and the quark is left with a net color of blue. The gluons at the far right, which have null quantum numbers, do not change the quark colors, and no gluons have any effect on quark flavors. At all times the baryon contains a red, a green and a blue quark; in the meson the color of the quark is balanced at each of the instants shown here by the anticolor of the antiquark. In practice it is not possible to determine the colors of the quarks; only the probability of each color can be calculated. In hadrons that are colorless the probabilities of the colors are equal.

do with angular momentum; it is determined by the number of particles included in a given group. Hypercharge is determined by the electric charges of those particles, and it is also related to both baryon number and strangeness. From the various possible combinations of the values that these two quantum numbers can assume it is possible to construct an array for each family of hadrons. These arrays, having in each case one, eight or 10 positions, predict the existence of all the known hadrons and of no others. The formation of these arrays can be described formally in the branch of mathematics called group theory. The arrays are said to be representations of the symmetry group SU(3), which stands for special unitary group for matrixes of size 3×3.

The quarks are also described by an SU(3) symmetry group. Gell-Mann designated the quarks by the arbitrary labels u, d and s, for "up," "down" and "sideways." All of them have the same spin angular momentum, 1/2 unit, and in the SU(3) group they form a family of their own; it is, obviously, a family of three. The three members of the quark family are distinguished by having different values of isotopic spin and hypercharge, and they differ in other quantum numbers as well. The electric charges assigned to them are particularly unusual. The u quark has a charge of $+2/3$, and the d quark and the s quark each have a charge of $-1/3$. The baryon numbers of the quarks are also fractional: all the quarks have a baryon number of $+1/3$. Strangeness, on the other hand, remains an integer quantum number; the u and d quarks have zero strangeness and the s quark has a strangeness of -1. For the corresponding antiquarks, whose symbols are $\bar{u}$, $\bar{d}$ and $\bar{s}$, the magnitude of each of these

quantum numbers is the same but the sign is changed.

The fundamental rule for the construction of hadrons from quarks is disarmingly simple: it states that all the quantum numbers of the hadron can be found by adding up the quantum numbers of the constituent quarks. The proton, for example, consists of two u quarks and one d quark, a configuration written uud. The electric charges are therefore $2/3 + 2/3 - 1/3$, for a total value of $+1$. The baryon number is $1/3 + 1/3 + 1/3$, or $+1$, and since the strangeness of each of these quarks is zero, the total is also zero. All the sums are in accord with the measured properties of the proton.

The positively charged pi meson is made up of a u quark and a $\bar{d}$ antiquark. The electric charges of the quarks are $+2/3$ and $+1/3$, for a total of $+1$, and the baryon numbers are $+1/3$ and $-1/3$, for the total baryon number of zero required of a meson. The strangeness again is zero. The spin-angular-momentum quantum number calls for a slightly more elaborate calculation because the spins of the quarks can be aligned in either of two ways, and the alignment determines the signs that must be supplied for the spin quantum numbers when they are summed. In all possible cases, however, combinations of three quarks or three antiquarks (baryons and antibaryons) must have half-integer spin, whereas combinations of a quark and an antiquark (mesons) must have integer spin.

The great strength of the quark model is that through this simple additive procedure the model correctly predicts all the quantum numbers of the known hadrons. In particular it should be pointed out that all allowed combinations of the quarks yield integer values of elec-

tric charge and baryon number, and that no other combinations do so (except in the trivial case of multiples of the allowed combinations). Furthermore, all the known hadrons can be constructed out of either three quarks or a quark and an antiquark.

In the past few years it has become apparent that there may be a fourth species of quark, bearing a new quantum number somewhat similar to strangeness and given the arbitrary name charm. The new quark (labeled c) adds another dimension to the symmetry group that describes the hadrons, and it predicts the existence of a multitude of new particles, some of which may already have been found. The addition of charm to the quark model, which seems increasingly to be justified by the experimental evidence, has a number of attractive features and can be considered to strengthen the model, but it has little effect on the problem of quark confinement.

In many respects quarks are like leptons. Both kinds of particle can apparently be represented as dimensionless points, and if they are without extension, they are also presumably without internal structure. All the quarks and all the leptons also share the property of having a spin of 1/2 unit. Finally, if the charm hypothesis is correct, then there are four members of each group: indeed, the appeal of that symmetry was one of the principal motivations for introducing the concept of charm. (On the other hand, four need not be the ultimate number of quarks and leptons. Both groups could have additional, undiscovered members.)

The resemblance between quarks and leptons is not a superficial one, but there are important differences between these two kinds of fundamental particle. In the first place, quarks participate in strong interactions, whereas leptons do not. The quarks also form aggregations of particles (hadrons), whereas there are no analogous composite structures made up of leptons. Why is it, however, that the quarks form only certain well-defined aggregations, those made up either of three quarks or of a quark and an antiquark? Many other combinations are conceivable—four quarks, two quarks, a quark and two antiquarks, even states made up of hundreds or thousands of quarks—but there is no evidence that any of them exist. A state of particular interest is the one represented by a solitary quark. Isolated leptons are commonplace; what distinguishing property of quarks prohibits them from appearing in isolation?

The concept that has provided the first, speculative answers to these questions was introduced in order to mend a conspicuous flaw in the quark theory.

		SPIN (J)	ELECTRIC CHARGE (Q)	BARYON NUMBER (B)	STRANGE-NESS (S)	CHARM (C)
QUARKS	u (UP)	½	+⅔	⅓	0	0
	d (DOWN)	½	−⅓	⅓	0	0
	s (STRANGE)	½	−⅓	⅓	−1	0
	c (CHARMED)	½	+⅔	⅓	0	+1
ANTIQUARKS	ū (UP)	½	−⅔	−⅓	0	0
	d̄ (DOWN)	½	+⅓	−⅓	0	0
	s̄ (STRANGE)	½	+⅓	−⅓	+1	0
	c̄ (CHARMED)	½	−⅔	−⅓	0	−1

PROPERTIES OF QUARKS are accounted for by assigning them quantum numbers, which can assume only certain discrete values. In the original quark model there were three kinds of quark, labeled u and d (for "up" and "down") and s (for "sideways" or "strange"). There is now evidence for a fourth quark species, labeled c (for "charm"). The quarks have fractional electric charge and fractional values of baryon number, a quantum number that distinguishes between two groups of particles. The spin quantum number reflects the intrinsic angular momentum of the quarks; the strangeness and charm quantum numbers recognize special properties of s and c quarks respectively. For each quark there is an antiquark with opposite quantum numbers.

The flaw involves an apparent conflict between the behavior of the quarks and one of their quantum numbers, spin angular momentum. The assignment of a spin of 1/2 unit to each of the quarks is essential if the spins of the hadrons are to be predicted correctly. Quantum mechanics, however, specifies rules for the behavior of particles with half-integer spins, and the quarks do not seem to obey them.

The quantum-mechanical rules postulate a connection between the spin of a particle and its "statistics," the set of rules that mandates how many identical particles can occupy a given state. Particles with integer spin are said to obey Bose-Einstein statistics, which allow an unlimited number of particles to be brought together in one state. Particles with half-integer spins obey Fermi-Dirac statistics, which require that no two identical particles occupy a state. This is the exclusion principle formulated by Wolfgang Pauli, the quantum-mechanical equivalent of the intuitive notion that no two things can be in the same place at the same time.

The most familiar application of Fermi-Dirac statistics and of the related exclusion principle is in atomic physics. There it governs the way electrons (which, being leptons, have a spin of 1/2) fill the orbitals, or energy levels, surrounding the nucleus. If an orbital contains one electron, one more can be added, provided that its spin is aligned in the direction opposite to that of the first electron's. With opposite spins the electrons do not have identical quantum numbers, and so they can occupy the same state, in this case an atomic orbital. Since there are only two possible directions for the spin, however, all other electrons are permanently excluded from the orbital.

The connection between spin and statistics is not well understood at a theoretical level, but it is not in doubt. In fact, formal proofs have been presented showing that the exclusion principle must be obeyed by all particles with half-integer spin, without exception. Like electrons, quarks move in orbitals, although their motion is measured not with respect to a nucleus but with respect to one another or to their common center of mass. For the least massive families of hadrons all the quarks should be in the same orbital: the smallest one. It follows that no two quarks in a hadron can have exactly the same quantum numbers.

In the quark model of the meson the requirements of Fermi-Dirac statistics can readily be accommodated. The two particles that compose a meson are a quark and an antiquark, and their quantum numbers are therefore different (in some cases they are exactly opposite). In the baryon, however, spin and statistics

QUARK FLAVORS		QUARK COLORS		
		RED	GREEN	BLUE
u (UP)		Q = +2/3 B = +1/3	Q = +2/3 B = +1/3	Q = +2/3 B = +1/3
		Q = 0 B = 0	Q = +1 B = 0	Q = +1 B = +1
d (DOWN)		Q = -1/3 B = +1/3	Q = -1/3 B = +1/3	Q = -1/3 B = +1/3
		Q = -1 B = 0	Q = 0 B = 0	Q = 0 B = +1
s (STRANGE)		Q = -1/3 B = +1/3	Q = -1/3 B = +1/3	Q = -1/3 B = +1/3
		Q = -1 B = 0	Q = 0 B = 0	Q = 0 B = +1
c (CHARM)		Q = +2/3 B = +1/3	Q = +2/3 B = +1/3	Q = +2/3 B = +1/3
		Q = 0 B = 0	Q = +1 B = 0	Q = +1 B = +1

ADDITIONAL QUANTUM NUMBER of quarks is called color, and it can assume three possible values, represented here by the primary colors red, green and blue. In contrast to color the original quark designations u, d, s and c are sometimes called quark flavors. (Both color and flavor are arbitrary terms; they do not have their usual meaning.) Each flavor of quark is assumed to exist in each of the three colors. In one model (*white boxes*) red, green and blue quarks of a given flavor are indistinguishable; they have the same values of electric charge (Q) and baryon number (B) and of all other quantum numbers. In an alternative theory (*gray boxes*) proposed by the author and M.-Y. Han quarks of different colors differ in electric charge and baryon number, and these quantum numbers can be given integer values. The Han-Nambu model cannot be excluded with certainty, but in this article fractional charges are assumed.

issue conflicting imperatives. In at least three baryons (*uuu, ddd* and *sss*) all three quarks have identical quantum numbers. Because there are three particles in the baryon, at least two of them must have their spins aligned the same way, and in many cases all three spins must point in the same direction. The exclusion principle seems to be violated.

A strategy for avoiding this uncomfortable conclusion was first proposed by O. W. Greenberg of the University of Maryland. Greenberg suggested that the quarks might not obey Fermi-Dirac statistics but could instead be governed by an unconventional set of rules he called para-Fermi statistics of order 3. Whereas in Fermi-Dirac statistics a state can be occupied by only one particle, in para-Fermi statistics it can be occupied by three particles, but no more.

Another approach to the problem was later suggested by M.-Y. Han of Duke University and me, and independently by A. Tavkhelidze of the Joint Institute for Nuclear Research in the U.S.S.R. and Y. Miyamoto of the Tokyo University of Education. Instead of changing the rules we changed the quarks. By assigning each quark an extra quantum number with three possible values, it is possible to arrange for all the quarks in a baryon to be of different species and therefore in different quantum-mechanical states. All that is necessary is some mechanism that will ensure that in every case each quark has a different value of

the new quantum number. This extra quantum number has come to be known as color. For the three values of the quantum number it is convenient to adopt the three primary colors red, green and blue; the antiquarks have anticolors, which can be represented by the complements of the primaries, respectively cyan, magenta and yellow. (None of these terms, of course, has any connection with its conventional meaning; they are arbitrary labels.)

The para-Fermi statistics can be regarded as a special case of the color hypothesis. The two theories are equivalent if the assumption is that color is completely unobservable. In that case quarks of different colors would appear to be identical in all their properties, and since there would be no way of telling one from another it would seem that identical quarks were obeying unconventional statistics. Color would be invisible, or, to put it another way, nature would be color-blind. The color hypothesis allows, however, for the possibility that color might be visible under some circumstances.

The introduction of color necessarily triples the number of quarks. If only the original three quarks are considered, then with color there is a total of nine; if the charmed quark is included, it too must have red, green and blue varieties, and the total number of quarks is 12. The number of hadrons, however, is not increased; the color hypothesis does not predict any new particles. The number of hadrons is unchanged because of the special way the colors are allotted to the quarks in a hadron. If the colors are to solve the problem of quark statistics, it is essential that a baryon contain one quark of each color; if a baryon could be made of three red quarks, for example, then the quantum numbers of all the quarks could well be identical. Only if all three colors are equally represented can obedience to the exclusion principle be ensured. Since we have assigned the quarks primary colors such a combination could be termed white, or colorless. As we shall see, the theory implies that all hadrons, both baryons and mesons, are colorless. The baryons are made up of equal quantities of red, green and blue; the mesons are equal mixtures of each color with its anticolor.

The formal treatment of the quark colors involves postulating another SU(3) symmetry group, exactly analogous to the one that determines the other properties of the quarks. The two quantum numbers that determine the quark colors are named, again by analogy to the original SU(3) group, color isotopic spin and color hypercharge. The properties determined by the original SU(3) symmetry are sometimes called quark flavors, and flavor, unlike color, is readily observed in experiments; it is, as it were, tasteable. The labels u, d, s and c represent the quark flavors, and they determine all the observable properties, such as electric charge, of the hadrons they compose. The symmetry among the flavors is not a perfect one, and quarks that differ in flavor have slightly different masses. Color, on the other hand, is an exact symmetry; in the usual formulation of the theory a quark of a given flavor has the same properties and the same mass regardless of its color.

Color was introduced into the quark theory as an ad hoc element to solve the problem of quark statistics. It has since become a central feature of the model. In particular it is thought to determine the forces that bind quarks together inside a hadron, and hence to have a profound influence on quark confinement. In this context the qualitative distinction between quarks and leptons begins to seem comprehensible. An important element in the distinction is that leptons do not form strongly bound states. If it is the color quantum number that is responsible for binding quarks together, then the absence of strong binding in leptons is readily understood, since the leptons do not have color.

In order to understand the forces between quarks it is helpful to consider first a more familiar force: electromagnetism. The electromagnetic force is described by Coulomb's law, which states that the force between two charged bodies declines as the square of the distance between them. For example, the force between a proton in the nucleus of an atom and one of the electrons surrounding the nucleus is described by this law. The force can be regarded as being transmitted by a field or by discrete particles: photons, the quanta of the electromagnetic field. Ultimately both the field and the force are derived from the electric charges of the particles; since those charges are unlike, the force is attractive.

The forces between quarks are in many respects similar, but they are somewhat more complicated. In the case of the electromagnetic field only one quantum number (electric charge)

PROPERTIES	CONSTITUENT QUARKS			HADRONS
	u	u	d	PROTON (p)
SPIN (J)	(½,½) +	(½,½) +	(½,−½) =	(½,½)
ELECTRIC CHARGE (Q)	⅔ +	⅔ −	⅓ =	+1
BARYON NUMBER (B)	⅓ +	⅓ +	⅓ =	+1
STRANGENESS (S)	0 +	0 +	0 =	0
CHARM (C)	0 +	0 +	0 =	0

		u	d̄	PION (π⁺)
SPIN (J)		(½,½) +	(½,−½) =	(0,0)
ELECTRIC CHARGE (Q)		⅔ +	⅓ =	+1
BARYON NUMBER (B)		⅓ −	⅓ =	0
STRANGENESS (S)		0 +	0 =	0
CHARM (C)		0 +	0 =	0

	ū	d̄	d̄	ANTINEUTRON (n̄)
SPIN (J)	(½,−½) +	(½,½) +	(½,½) =	(½,½)
ELECTRIC CHARGE (Q)	−⅔ +	⅓ +	⅓ =	0
BARYON NUMBER (B)	−⅓ −	⅓ −	⅓ =	−1
STRANGENESS (S)	0 +	0 +	0 =	0
CHARM (C)	0 +	0 +	0 =	0

	u	d	s	LAMBDA (Λ°)
SPIN (J)	(½,½) +	(½,−½) +	(½,½) =	(½,½)
ELECTRIC CHARGE (Q)	⅔ −	⅓ −	⅓ =	0
BARYON NUMBER (B)	⅓ +	⅓ +	⅓ =	+1
STRANGENESS (S)	0 +	0 −	1 =	−1
CHARM (C)	0 +	0 +	0 =	0

		c	ū	CHARMED MESON (D°)
SPIN (J)		(½,½) +	(½,−½) =	(0,0)
ELECTRIC CHARGE (Q)		⅔ −	⅔ =	0
BARYON NUMBER (B)		⅓ −	⅓ =	0
STRANGENESS (S)		0 +	0 =	0
CHARM (C)		1 +	0 =	+1

QUARKS COMBINE to make up the class of observed particles called hadrons. Two kinds of quark combination are possible. In one of them three quarks bind together to form a baryon (such as the proton) or three antiquarks bind to form an antibaryon (such as the antineutron). In the other kind of quark binding a quark and an antiquark make up a meson (such as the pion). The properties of these hadrons are determined by the simple rule that the quantum numbers of the hadron are the sums of the quark quantum numbers. All the allowed combinations of quarks give integer values of electric charge. The baryon numbers add in such a way that all baryons have a value of +1, antibaryons −1 and mesons 0. Strange particles, such as the lambda baryon, are those that have at least one s quark; charmed particles have at least one c quark. The spin quantum number is a vector and requires a more complicated arithmetic, but the result of the addition is that all baryons and antibaryons have half-integer spin and all mesons have integer spin. The great success of the quark theory is that all the allowed combinations yield known hadrons and no other combinations do so. The problem of quark confinement is why only these combinations should exist, and why solitary quarks are not observed.

is involved in generating the field; the fields between quarks are generated by two quantum numbers: color isotopic spin and color hypercharge. To continue the analogy with electromagnetism, these quantum numbers can be regarded as two varieties of "color charge."

If a combination of quarks is to be stable, it is obvious that the forces between them must be attractive. That can be arranged by ensuring that the quarks in a baryon, for example, are all of different colors, since the quarks will then have unlike values of the two kinds of color charge. A red quark and a green quark will be bound together because their color-isotopic-spin numbers are of opposite sign; the blue quark will be bound to both of the others because of a difference in the sign of the color-hypercharge quantum number. A similar mechanism generates an attractive force between a quark of one color and an antiquark of the corresponding anticolor, as in a meson. The forces favor just those combinations that have been identified as white, or colorless.

Actually the situation is still more complicated. Whereas the electromagnetic force is transmitted by a single kind of particle, the photon, the force associated with quark colors requires eight fields and eight intermediary particles. These particles have been named gluons because they glue the quarks together. Like the photon, all of them are massless and have a spin of 1; like the quarks themselves, they have not been detected as free particles.

The eight gluons can be regarded as having composite colors, made up of the various combinations of three colors and three anticolors. All together there are nine such combinations, but one of them receives equal contributions from red combined with antired, green combined with antigreen and blue combined with antiblue. Since that combination is effectively colorless it is a trivial case and is excluded, leaving eight colored gluons.

Quarks interact by the exchange of gluons; when they do so, they can change their colors but not their flavors. At all times a baryon contains red, green and blue quarks, but because gluons are continually exchanged it is not possible to say at any given moment which quark is which color. Similarly, a meson always consists of a quark of one color and an antiquark of the complementary anticolor, but the three possible combinations of color and anticolor are equally represented. In quantum mechanics we can have no certain knowledge of quark colors; rather, we can know only the probability that a quark is a given color. If all hadrons are colorless, the probabilities for the three colors are equal.

The model of hadrons in which these

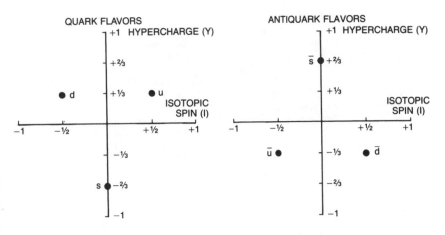

CLASSIFICATION OF QUARKS is governed by a fundamental symmetry of nature. The quark flavors are determined by two quantum numbers, isotopic spin and hypercharge; each quark or antiquark represents a unique combination of these numbers. Color is determined by two other quantum numbers, named by analogy color isotopic spin and color hypercharge. Both flavor and color can be described by means of the concept of a symmetry group. Flavor is a "broken symmetry" because quarks with different flavors differ in properties such as mass and electric charge. Color is an exact symmetry: two quarks of the same flavor but of different colors differ only in their colors and are otherwise indistinguishable. The quark colors are thought to generate forces binding quarks together. These forces result from two kinds of field, or color charge, associated with quantum numbers color isotopic spin and color hypercharge.

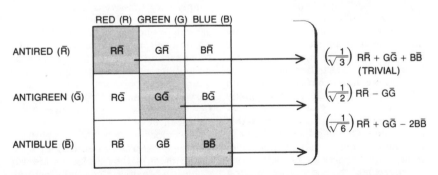

COLORED GLUONS are the particles that transmit forces between colored quarks; they are the quanta of the fields generated by the color quantum numbers just as the photon is the quantum of the electromagnetic field. The gluons can be regarded as combinations of color and anticolor. In the first analysis there are nine possible combinations of three colors and three anticolors; six of these combinations (white squares) are straightforward but three require special treatment (gray squares). The three involve combinations of a single color with the corresponding anticolor, and in each combination all quantum numbers cancel out. States with null quantum numbers can be combined at will, but only three of these combinations of combinations need be considered. One is $R\bar{R} + G\bar{G} + B\bar{B}$, in which the quantum numbers are still null; the case is trivial and can be eliminated. The remaining two are $R\bar{R} - G\bar{G}$ and $R\bar{R} + G\bar{G} - 2B\bar{B}$, which can be treated like the other six gluons except that numerical correction factors are required.

particles are composed of quarks bound together by the exchange of gluons can be given an elegant mathematical formulation. The model is an example of a non-Abelian gauge theory, a kind of theory invented by C. N. Yang of the State University of New York at Stony Brook and Robert L. Mills of Ohio State University. A gauge theory is one modeled on the theory of electromagnetism developed by James Clerk Maxwell. A characteristic of such theories is that any particle carrying a given quantum number, or charge, generates a long-range field whose strength is proportional to the quantum number. In Maxwell's theory the quantum number in question is electric charge; in the model of hadron structure there are two such numbers, those associated with the quark colors.

Maxwell's theory is an Abelian gauge theory; non-Abelian gauge theories are distinguished from it by the fact that the fields themselves carry quantum numbers. A field can therefore act as a source of itself. Einstein's theory of gravitation is also a non-Abelian gauge theory in that the gravitational field itself generates gravity. Electromagnetism and the weak force, which is responsible for certain kinds of radioactive decay, have recently been combined in another non-Abelian gauge theory by Steven Weinberg of Harvard University and by Abdus Salam of the International Centre for Theoretical Physics in Trieste. The colored-quark model could now provide a similar framework for understanding the strong force. These four forces—strong, weak, electromagnetic and gravitational—are the only ones known in nature. It would bring great aesthetic satisfaction if all four could be understood through the same kind of theory.

Before describing schemes for the confinement of quarks it would be well to consider the possibility that they are not confined at all. Perhaps they have been there all the time but we have not been able to detect their presence, or perhaps we have confused them with some ordinary particle. If a fractionally charged quark could escape from a hadron, it would almost certainly be stable in isolation. One quark might decay to yield another quark, perhaps along with some ordinary particles, but at least one quark species—the one with the smallest mass—must be stable. It could not decay because all particles other than quarks have integer charges, which cannot be created from the decay of a fractionally charged quark.

Such free, stable quarks could well come to rest among the atoms of ordinary matter. Similarly, if they can escape, they should be found in the debris produced by high-energy collisions of hadrons, both in particle accelerators and when cosmic rays collide with atoms in the atmosphere. The principal argument against the existence of such free quarks is that they have not been found in ordinary matter, even in minute concentrations, and they have not been seen in the aftermath of hadron collisions.

If fractionally charged quarks were present, they could readily be detected and recognized. Charged particles are detected by the ionization they cause in the atoms surrounding them. The extent of the ionization is proportional to the square of the particle's electric charge. Thus a quark with a charge of 1/3 would produce only one-ninth the ionization of a particle with a charge of 1, and it could easily be distinguished from ordinary particles.

Perhaps, however, the quarks do not have fractional charges. With the addition of color to the quark theory it becomes possible to assign each quark integer values of both electric charge and baryon number, and Han and I proposed such a model in 1965. The model has the effect of making color visible, in the sense that quarks of different colors carry different masses, electric charges and baryon numbers and can therefore be distinguished. For each flavor of quark all the electric charge $(+1$ or $-1)$ would be assigned to one color, and the other two colors would have zero charge. If all colors must be represented equally, then the total charge of the hadron would be correct. If the quarks do have integer charges, then a free quark in the laboratory would not look much different from an ordinary baryon and might easily be misidentified. This possibility cannot yet be excluded with certainty.

Another hypothesis suggests that it is difficult to extract quarks from hadrons, but not impossible. Perhaps they are simply very massive and the accelerators operating today are not powerful enough to liberate them. That hypothesis, however, requires that the mass of a free quark and that of a bound quark be quite different. Indeed, a single isolated quark might be more massive than a baryon composed of three quarks, a notion that is difficult to understand if it is not inconceivable.

The color theory of hadron construction leads naturally to at least a partial confinement of the quarks. An atom is stablest when it is electrically neutral, that is, when it has attracted just enough electrons to balance the positive charge of the nucleus. Any attempt to add an extra electron or to remove one of those already bound is resisted. In the same way a system of quarks is stablest when all three colors, or a color and an anticolor, are present; then the hadron is neutral with respect to the two kinds of color charge. This result is hardly surprising: the color quantum numbers were introduced precisely in order to achieve an equal representation of the colors in baryons. It follows that since an isolated quark is necessarily a colored particle, it is an energetically unfavorable configuration. Free quarks will tend to associate to form colorless hadrons, just as free electrons and ionized atoms will tend to recombine. This aspect of the quark colors does not exclude the possibility of there being free quarks, but it does strongly inhibit their formation. It requires that a free quark or any other colored state be less stable, or more massive, than colorless states.

The quark model has changed significantly and grown far more elaborate since it was proposed in 1963. There is every reason to believe it will go on evolving, and it is entirely possible the present perceived need to explain the confinement of quarks will be altered by subsequent events—including, perhaps, the discovery of a free quark. The fact remains that experimental searches for quarks, guided by reasonable conjectures about their properties, have failed to reveal their presence. The consistently negative results demand explanation. One approach is to postulate a mechanism that permanently confines the quarks to the interior of a hadron, so that free quarks are not merely discouraged but are absolutely prohibited. Several theories can provide such a mechanism, and some of them exhibit exceptional ingenuity.

One of these ideas grows directly out of the underlying gauge theory of the interactions between colored quarks. Once again the principle can be effectively illustrated by considering first the analogous phenomena observed in electromagnetic interactions of matter.

The inverse-square relation of Coulomb's law has been verified with great precision at large distances, but it is not valid when the force between charged particles, such as electrons, is measured at extremely short range. The discrepancy is caused by the spatial distribution of the electron's charge. At the core of the electron there is a negative charge, called the bare charge, of very large magnitude; indeed, it may well be infinite. This charge induces in the vacuum surrounding it a halo of positive charge, which almost cancels the bare charge. The effective charge of the electron, when measured from a distance, is simply the difference between these two charges. A test particle able to approach the electron at close range will penetrate the shielding of positive charge and will begin to perceive the large bare charge.

Electromagnetism, it will be remembered, is an Abelian gauge theory, whereas the colored-quark theory of the strong force is a non-Abelian one. In this context the distinction is a crucial one, a fact that has been demonstrated by H.

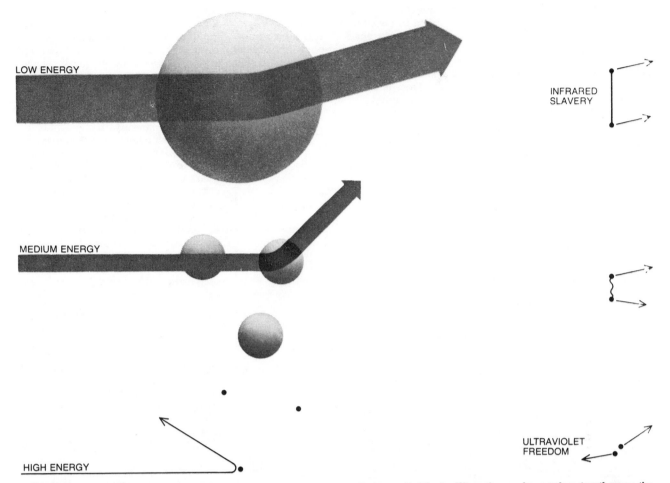

LOW ENERGY

MEDIUM ENERGY

HIGH ENERGY

INFRARED
SLAVERY

ULTRAVIOLET
FREEDOM

INFRARED SLAVERY, one proposed explanation of quark confinement, is a concept derived directly from the field theory that describes interactions of colored quarks and gluons. The theory holds that quarks at large distances (about the size of a hadron, 10^{-13} centimeter) are bound tightly together and move in concert. The long-distance behavior of the quarks is examined by low-energy probes of the hadron, and indeed at these energies the hadron appears to be a cohesive, unified body. When the quarks are close together, on the other hand, they are only weakly bound and can move independently. The forces between quarks at this range are investigated by high-energy probes, and such experiments have observed centers of mass in the hadron that seem to move freely. The names infrared slavery and ultraviolet freedom were applied to these phenomena through analogy with the relative energies of infrared and ultraviolet radiation.

David Politzer of Harvard and David Gross and Frank Wilczek of Princeton University. In the non-Abelian theory the bare charge does not induce a shielding charge but an "antishielding" one. Thus a quark with a color charge induces around it additional charges of the same polarity. As a result the color charge of the quark is smallest at close range; as a particle recedes from the quark the charge gets larger. The corresponding force law is dramatically different from Coulomb's law: as the distance separating two color-charged particles increases, the force between them could remain constant or could even increase.

A test particle colliding with a hadron at high energy inspects the behavior of the constituent quarks over very small distances and during a very brief interval. This fact is established mathematically by the uncertainty principle, which relates the time and distance in which a measurement is performed to the energy and momentum of the test particle. It

can be understood intuitively by remembering that a high-energy particle moves at nearly the speed of light and that it "sees" the quarks for only a brief moment, during which they can move only a short distance. The non-Abelian gauge theory predicts that such a high-energy probe will reveal the quarks to be essentially free particles, moving independently of one another, since at small distances the color charge declines and the quarks are only loosely bound together.

A low-energy investigation of a hadron, on the other hand, should see quarks that are rigidly tied together and therefore move as a unit. At these comparatively low energies the quarks are being observed during a more extended period, and they can interact over greater distances. Hence the more powerful long-range effects of the color gauge fields grip the quarks and bind them to one another.

Since the theory is a non-Abelian one, the gluons are subject to the same con-

straints as the quarks, and they are confined just as efficiently. The gluons, or the fields they represent, generate fields of their own that have the same character as the color fields of the quarks. The resulting behavior of the gluons contrasts sharply with that of photons, the quanta of the gauge fields in the Abelian theory of electromagnetism. Photons do not themselves give rise to an electromagnetic field, and they escape from such a field without hindrance.

These two opposing aspects of the color gauge theory have been given the picturesque names infrared slavery and ultraviolet freedom. The terms do not refer to those particular regions of the electromagnetic spectrum but are simply intended to suggest low-energy and high-energy phenomena respectively. Ultraviolet freedom is also known as asymptotic freedom, because the state of completely independent movement is approached asymptotically and never actually achieved. The effect may have been observed in collisions of electrons with protons, where it has been found that at very high energy the proton behaves as if it were a collection of free quarks.

The concept of infrared slavery provides an obvious means of explaining the confinement of quarks. If the effective color charge continues to increase indefinitely with increasing distance, then so does the energy needed to pull two quarks apart. Achieving a macroscopic separation would require an enormous input of energy and would surely be a practical impossibility.

The spatial distribution of the color charge is not known, however, for macroscopic distances; indeed, nothing is known of it for any distance greater than the approximate size of a hadron: 10^{-13} centimeter. Whether or not infrared slavery can account for quark confinement depends on the details of the charge distribution. It should be pointed out, however, that the charge need not increase indefinitely to entrap the quarks permanently. It need only increase to the point where the energy required to further separate the quarks is equal to the energy needed to create a quark and an antiquark. When that energy is reached, the quark-antiquark pair can materialize. The newly created quark replaces the one extracted, and the antiquark binds to the displaced quark, forming a meson. The result is that a quark is removed from the hadron but is not set free; all we can observe is the creation of the meson.

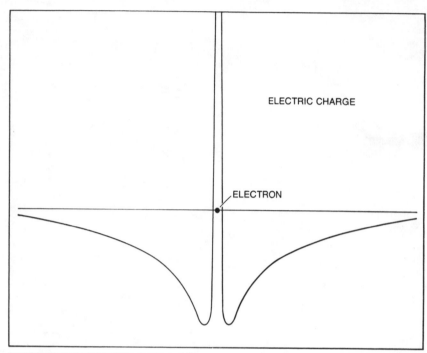

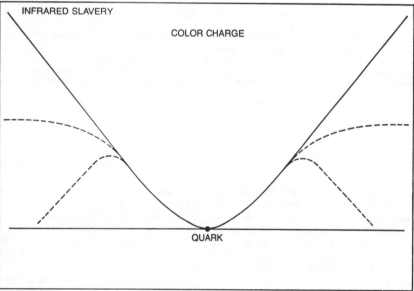

A kind of charge that increases with distance and a force that remains constant with distance seem to contradict an intuitive sense of how matter ought to behave. Quantum mechanics has contradicted intuition before, and made no apology for it, but in this case an explanation, and even a pictorial representation, of how the effects might arise may be possible. This explanation is a feature of another model of quark confinement, called the string model.

The string model grew out of mathematical formulas introduced by Gabriele Veneziano of the Weizmann Institute of Science. In the model hadrons are regarded as flexible, extensible strings in rapid rotation. The string is massless, at least in that it has no material "beads" along its length, although it does have potential and kinetic energy. The string is given a certain fixed tension as one of its intrinsic properties, so that the ends of the string tend to pull toward each other with constant force. The tension represents potential energy (just as the tension of a stretched spring does), and the magnitude of that energy is exactly proportional to the length of the string. If the string were stationary, its intrinsic tension would cause it to collapse, but the system can be kept in equilibrium by spinning the string. As the string spins it stretches, and when its length is such that its ends move with the speed of light, the centrifugal force balances the tension. (The ends are allowed to move with the speed of light, and indeed are required to do so since they are massless.)

Because of the relations between length, energy and rotation that have been built into the string model, the angular momentum of the system is proportional to the square of the total energy. In this respect the model reflects an important observed property of hadrons: When the angular momentum of hadrons is plotted against the square of

DISTRIBUTION OF THE COLOR CHARGE might explain the effects of infrared slavery and ultraviolet freedom. The distribution seems to be quite different from that of the more familiar electric charge. The electron has at its core a large and possibly infinite negative charge, called the bare charge, which induces in the vacuum surrounding it a positive charge of almost equal magnitude; the effective charge of the electron observed at a distance is the difference between these charges. The bare color charge, in contrast, is thought to be very small and possibly zero, but it induces a surrounding charge of the same polarity, so that the effective charge increases, perhaps without limit, as it spreads out in space. From these charge distributions it follows that electrically charged particles obey Coulomb's law: the force between them declines as the square of the distance. Particles bearing a color charge, on the other hand, obey a very different law: the force between them remains constant, regardless of distance, and the energy with which they are bound together (or the energy that must be supplied in order to pull them apart) increases with distance. The actual distribution of the color charge has not been measured at large distances, so that several continuations of graph are possible (*broken lines*).

their mass or energy, the result is a series of parallel lines, named Regge trajectories after the Italian physicist Tullio Regge. The relation between angular momentum and energy embodied in the string model provides a possible explanation for the observation that all the Regge trajectories are straight lines.

Quarks can be incorporated into the string model simply by attaching them to the ends of the strings. The quarks are then assumed to carry the quantum numbers of the hadron while the string carries most of the energy and momentum. Quark confinement follows as a natural consequence of the properties of the string. It is assumed that the quarks cannot be pulled off, and so the only way they can be separated is by stretching the string. Any increase in the length of the string, however, demands a proportionate increase in its energy, so that once again large separations are impossible. Nevertheless, even if the string cannot be stretched without an inordinate supply of energy, it might be snapped in two. At the breaking point a newly created quark and antiquark would be welded to the broken ends, with the result that a meson would be created. In all these interactions the string model can be seen to give results equivalent to those of the infrared-slavery hypothesis, even though the underlying description of the hadron has a quite different form.

What is the stuff of the massless spinning string? One appealing interpretation has been proposed by Holger B. Nielsen and P. Olesen of the Niels Bohr Institute in Denmark; in explaining it we shall return again to the consideration of electromagnetism. Coulomb's law describes an electromagnetic field in three-dimensional space, and if the field is represented by discrete lines of force, it is apparent that the strength of the field declines with distance because the lines spread out in space. Their density decreases as the square of the distance, giving the familiar force law. If all the lines of force could be compressed into a thin tube, the lines could not spread out and the force would remain constant, regardless of the distance.

The distinctive geometry of the string suggests that it might be regarded as such a one-dimensional gauge field. The properties of the string itself—in particular the inherent tensile force and the variation of energy with length—are then as predicted by the model. Furthermore, the bizarre properties of the color gauge field are given a simple and intuitively appealing explanation. It is no longer the force itself that is peculiar; the force is a conventional one, obeying the same kind of law as electromagnetism. The peculiar properties all derive from the geometry imposed on the field.

Fields that are virtually one-dimensional can actually be created on a macroscopic scale. If a superconductor (an

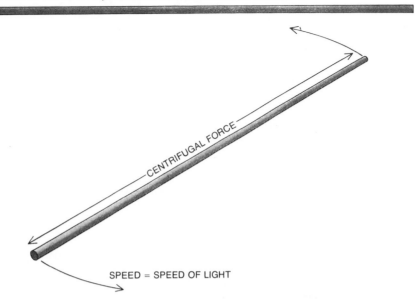

STRING MODEL of hadron structure leads to another possible explanation of quark confinement. The model assumes that a hadron is made of a massless, one-dimensional string that has as one of its intrinsic properties a constant tension per unit length. Because of its tension the string tends to collapse, but it can be kept in equilibrium by centrifugal force if it is made to spin so that its ends move with exactly the speed of light. These properties of the string imply that its energy is proportional to its length and that its angular momentum is proportional to the square of its energy, a relation that has been experimentally verified for the hadrons.

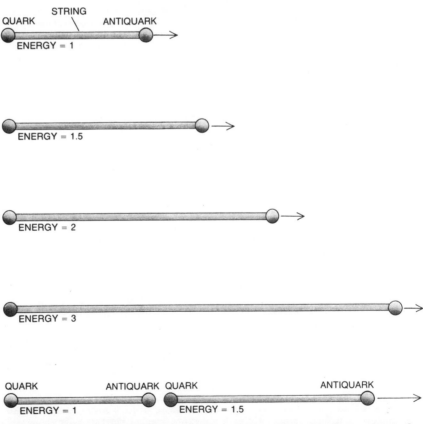

QUARKS WELDED TO STRINGS might be effectively confined. In order to separate the quarks it is necessary to stretch the string, but since the energy of the string is proportional to its length the energy required to pull the quarks apart increases in proportion to the separation. A macroscopic separation could be obtained only at the cost of enormous energy. In fact, isolation of a quark might not be possible at any energy, since as soon as enough energy had been supplied to create a quark and an antiquark the string might snap and these new particles appear at the ends. Thus the result is not the liberation of a quark but the creation of a meson.

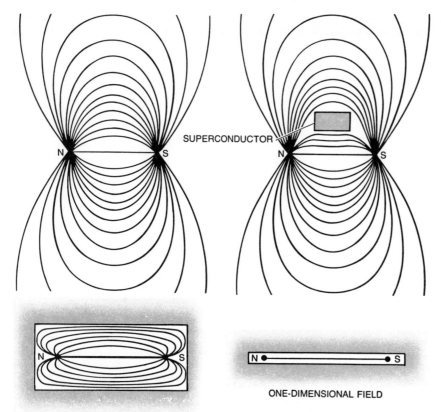

GEOMETRY OF THE STRING might be explained through an analogy with the behavior of a magnetic field in the vicinity of a superconductor. The strength of a magnetic field declines as the square of the distance because the lines of force spread out in three-dimensional space. The flux lines are expelled by a superconductor, and if two magnetic poles are surrounded by a superconducting medium, the field is confined to a thin tube. Under these circumstances the force between the poles is constant and the energy required to pull them apart increases linearly with their separation. A string might be a similar one-dimensional field, confined not by a superconducting medium but by the vacuum. Quark confinement could then be explained even if the color charge does not increase with distance but obeys a law like that of the electric charge.

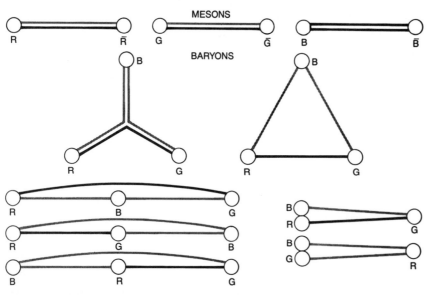

CONFIGURATION OF STRINGS linking quarks is not obvious in all cases and represents a serious impediment to the further development of the string model. It is convenient to consider two kinds of string, one associated with each of the color quantum numbers, color isotopic spin (*black*) and color hypercharge (*gray*). The binding together of a quark and an antiquark in a meson by these strings is straightforward, but baryons demand a more complex structure, for which there are several alternatives. The baryon might resonate between the various possible structures, but not all of them are satisfactory. The quarks must be able to interchange their colors without altering the mass or other properties of the hadron, but that condition is not invariably satisfied. Moreover, quark colors actually give rise to eight fields, associated with the eight gluons, rather than to two, and there is no obvious way of incorporating all into the model.

electrical conductor cooled to the superconducting state) is put into a magnetic field, the lines of force are expelled from the superconducting medium. If the two poles of a magnet are completely immersed in a superconductor, the lines of force are confined to a thin tube between the poles, where the superconductivity is destroyed. The tube of flux lines carries a fixed amount of magnetic energy per unit length, and the amount of magnetic flux is quantized. An exact analogy requires only that we assume that the effects of a superconducting medium on the magnetic field are duplicated in the effects of the vacuum on the color gauge field. A theory based on this comparison has been described mathematically; in it the quarks are likened to the hypothetical carriers of magnetic charge, the magnetic monopoles.

The string is a novel and amusing model of hadron structure, but attempts to make a complete and quantitative theory of it have encountered difficulties. The placement of the quarks at the ends of the string is rather arbitrary. This raises no serious problems in the case of the meson, which can be regarded as a single hank of string with a quark and an antiquark at the ends, but it is not clear what structure should be assigned to the baryon. Several configurations are possible, such as a three-pointed star, or a triangle with a quark at each vertex. The relation of mass or energy to angular momentum is very similar in baryons and mesons (that is, their Regge trajectories are nearly parallel), which implies that the internal dynamics of the two kinds of particle are also similar. This observation favors yet another possible baryon structure: a single string with a quark at one end and two quarks at the other end. In such a model, however, the colors can be assigned to the quarks in three ways, which are not equivalent to one another. Perhaps the baryon resonates between these configurations, much as the benzene ring resonates between its various possible structures.

The color-isotopic-spin and color-hypercharge quantum numbers can be accommodated in the string model by assuming that there are two kinds of string, each carrying the field associated with one of the quantum numbers. All together, though, there are eight gauge fields, represented by the eight color-anticolor combinations of the gluons. Are there then eight kinds of string also? How are we to describe the changes in the quark colors resulting from the emission or absorption of a gluon? These questions have yet to be answered satisfactorily. It may be that the simple, pictorial character of the string model is too naïve for a system in which quantum-mechanical effects are essential.

The third major attempt to account for the confinement of quarks takes a somewhat different approach but reaches a similar conclusion. This model has been proposed by Kenneth A. Johnson of the Massachusetts Institute of Technology and by others. It takes as one of its given initial conditions that the quarks are confined and from that assumption attempts to calculate known properties of the hadrons.

To provide containment the model employs what is perhaps the most obvious device: the quarks are trapped inside a bag, or bubble. It is a feature of the model that the quarks cannot penetrate the fabric of the bag, but by exerting pressure from inside they can inflate it. The energy of the bag itself, however, is proportional to its volume, so that large and potentially unlimited amounts of energy are required to separate the quarks. The system reaches equilibrium when the bag's tendency to shrink, in order to minimize its energy, is balanced by the pressure of the quarks inside, which move freely like the molecules of a gas. Interactions of the quarks inside the bag are governed by the standard non-Abelian gauge theory.

From the bag model it is possible to compute various properties of the proton and the neutron and of other hadrons with reasonable accuracy. The model is not very different in spirit from the Nielsen-Olesen description of the string. In one case the critical relation is between length and energy, in the other it is between volume and energy, but the effect is the same. The bag might be regarded as a string that is as thick as it is long. Conversely, if a round bag is spun fast enough, it elongates, that is, it turns into a string. Perhaps the bag will prove to be the appropriate model for discussing the ground states of hadrons, and the string will be applied to their excited, rotating states.

Each of these three models achieves its objective: it provides a mechanism for sequestering quarks inside hadrons. Each model can also account for a few properties of hadrons, but none can be considered definitive. Perhaps the comprehensive theory that will ultimately emerge will combine features of several models; for example, it would be useful to have the concept of ultraviolet freedom in the string models.

One approach to such a synthesis is being attempted by Kenneth G. Wilson of Cornell University. In Wilson's model the continuous space-time of the real world is approximated by a lattice in which the cells are the size of a hadron. Quarks can occupy any of the lattice sites and the color gluon fields propagate along straight lines (strings) linking them. Quark confinement is automatic.

Quarks are a product of theoretical reasoning. They were invented at a time when there was no direct evidence of their existence. The charm hypothesis added an extra quark explaining the properties of another large family of particles when those particles had themselves never been seen. Color, a concept of even greater abstraction, postulates three varieties of quark that may be distinct but completely indistinguishable. Now theories of quark confinement suggest that all quarks may be permanently inaccessible and invisible. The very successes of the quark model lead us back to the question of the reality of quarks. If a particle cannot be isolated or observed, even in theory, how will we ever be able to know that it exists?

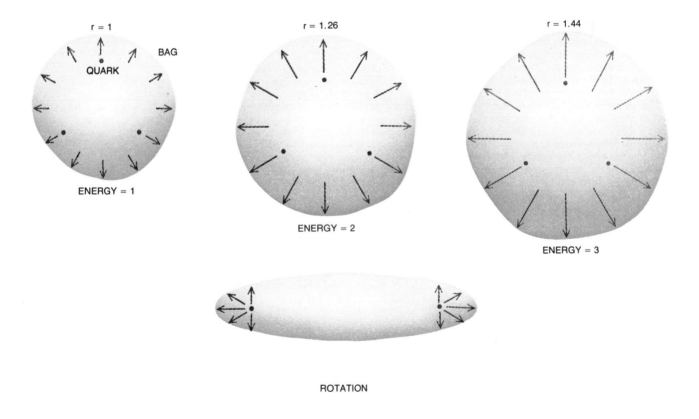

BAG MODEL of hadron structure offers a third mechanism for confining quarks. Indeed, in this model confinement is one of the initial assumptions: the quarks are assumed to be trapped inside a bag whose surface they cannot penetrate. The bag is kept inflated by the pressure of the quarks inside it, much as a balloon is inflated by the pressure of the gas inside it. The quarks can be separated only by increasing this inflation. The energy of the bag itself, however, is proportional to its volume, so that every increase in the distance between the quarks requires an additional application of energy. The bag model and the string model are closely related, and the connection between them becomes obvious when the bag is rotated rapidly: it then elongates to form an object essentially indistinguishable from a string.

8

The Bag Model
of Quark Confinement

by Kenneth A. Johnson
July, 1979

Quarks appear to be real, and yet they have not been observed in isolation. One hypothesis for why they have not been is that they are confined in bags analogous to the bubbles in a liquid

Subatomic particles are classified in two broad categories: the leptons (the electron, the muon, the tau particle and the neutrinos associated with each of them), which do not interact strongly and appear to have no constituent parts, and the hadrons (including the proton, the neutron and the pion), which do interact strongly and show signs of an inner structure. The force that holds neutrons and protons together in the atomic nucleus is one manifestation of the strong interaction between hadrons. That interaction is one of the four forces of physics, the others being the electromagnetic force, the weak force and gravitation.

More than 200 kinds of hadron have been identified over the past few decades. They too have been divided into two categories, depending on how they decay. The baryons are particles that decay ultimately into the proton, and the mesons are particles that decay entirely into leptons and photons (quanta of electromagnetic energy) or into proton-antiproton pairs.

The hadrons all appear to be combinations of the constituents named quarks. The quark model amounts to an impressive simplification of nature because at present just five "flavors," or kinds, of quark (and in most cases just three flavors) can account for the properties of the multitude of hadrons. The strong interaction binds the quarks together to form baryons and mesons. On the quark model baryons are made up of three quarks, and mesons are made up of a single quark-antiquark pair. What is intriguing about quarks, whose flavors are fancifully labeled "up," "down," "strange," "charmed" and "bottom," is that none has ever been observed in isolation, in spite of many attempts to find one. To explain this phenomenon I and a group of co-workers (Allen Chodos, Robert L. Jaffe, Charles B. Thorn and Victor F. Weisskopf) at the Massachusetts Institute of Technology have developed a theoretical model in which the

quarks making up a hadron are imprisoned in what may be called a bag or bubble. The model serves to clarify many aspects of the strong interaction between quarks.

When Murray Gell-Mann and George Zweig of the California Institute of Technology introduced the quark model in 1963, quarks were not universally regarded as real entities that might be found in the laboratory. The possibility was considered that they might just be theoretical constructs for bringing order to the unwieldy collection of hadrons. The quarks were thought to be peculiar objects because they carry only a fraction ($+2/3$ or $-1/3$) of the electric charge of the proton, whereas all other known physical entities carry an integral multiple of that charge. Moreover, the quarks seemed to violate the exclusion principle, the quantum-mechanical equivalent of the intuitive notion that no two things can be in the same place at the same time.

In the late 1960's the status of quarks as the fundamental constituents of all hadrons was dramatically confirmed by a series of experiments done by groups led by Jerome I. Friedman and Henry W. Kendall of M.I.T. and by Richard E. Taylor of the Stanford Linear Accelerator Center (SLAC). High-energy electrons were inelastically scattered off

protons, and the scattered distribution indicated the existence of particles within the protons. These particles behaved the way quarks were posited to behave and so they were identified as quarks. The scattering experiment demonstrated not only the physical reality of quarks but also their pointlike nature. Therefore all the fundamental constituents of matter, quarks as well as leptons, seem to act like dimensionless points.

The fact that three quarks make up a proton accounts for their fractional charge (the proton being composed of two "up" quarks, each with a charge of $+2/3$, and one "down" quark, with a charge of $-1/3$). Moreover, the introduction of the new quantum number, or physical property, labeled "color" makes it possible to reconcile the behavior of the quarks with the exclusion principle. Yet since no quark has ever been seen in the free state, quarks are still peculiar objects. It is relatively easy to remove an atom from a molecule, an electron from an atom or a proton from an atomic nucleus. Only a relatively small amount of energy is needed to free any one of these constituent particles. Huge amounts of energy have not succeeded in removing even one quark from one hadron.

At first some physicists believed a

LINES OF ELECTRIC FORCE (*top*) generated by opposite electric charges are curves in space indicating the direction of the electric force a positive charge would be subjected to if it were placed on the curves. The number of lines in a given spatial region indicates the magnitude of the force; the greater the number of lines, the stronger the force. Electric-field lines fill all space, but at a distance from the charges the field-line density is low and therefore the force is too. Lines of "color" force (*bottom*) link the quark-antiquark pair of a typical hadron: any subatomic particle that is subject to the interaction known as the strong force and that shows signs of an inner structure. Color is the designation of a property of quarks that has nothing to do with visual color. Each kind of quark comes in three colors: "red," "blue" and "yellow." The antiquarks have anticolors: "antired," "antiblue" and "antiyellow." All hadrons are "white," that is, they are colorless averages of the three colors and three anticolors. Two-quark hadrons consist of a quark of one color, say red (R), and the antiquark of the corresponding anticolor, say antired ($\bar{R}$). In the bag model of hadrons quarks are confined to a bag or bubble within a hadron, and so the color-field lines are also confined. The fact that the lines are parallel intensifies the force acting between the charges because the lines all pull in the same direction.

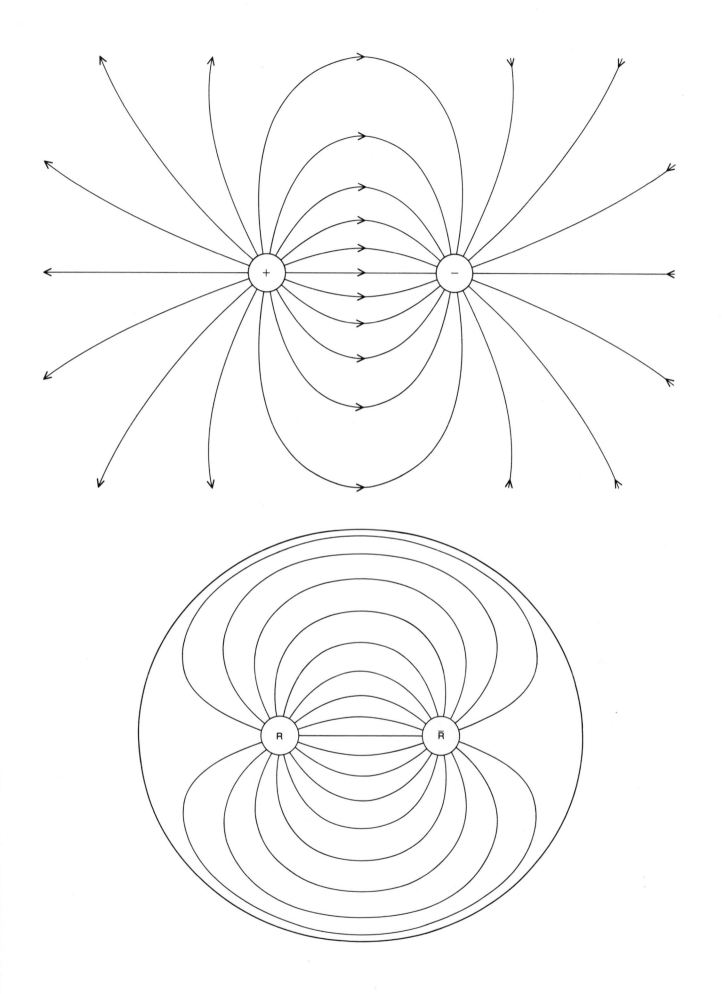

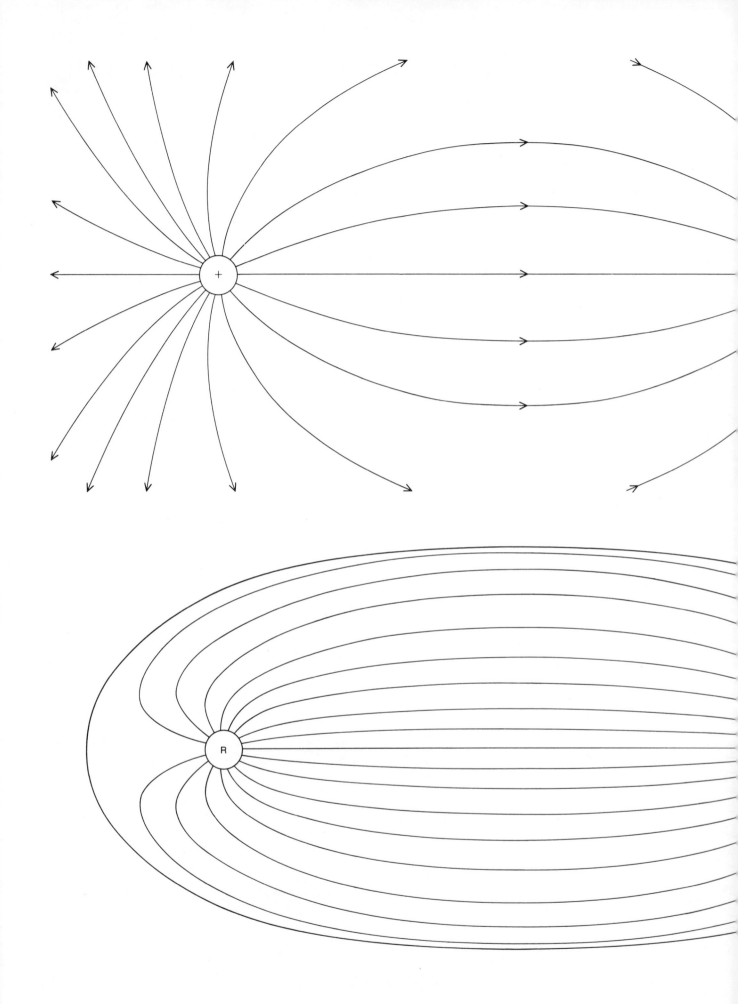

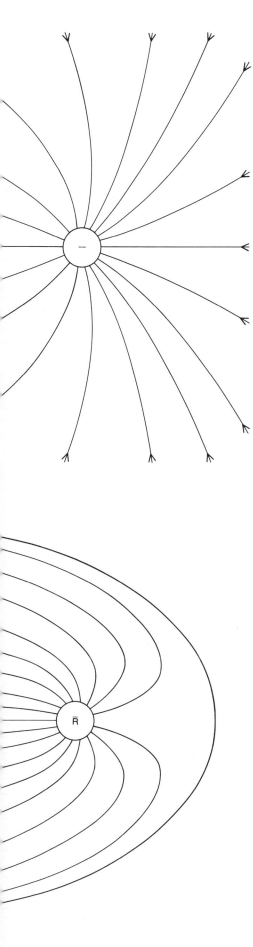

quark would be liberated as soon as particle accelerators were upgraded to higher energies. That has not happened, and quarks are now thought to be permanently confined to the interior of hadrons. In the past decade theorists have proposed several models of quark confinement; the bag model is only one of them. The bag model is gaining adherents because it may be closely related to what most theorists consider to be the best candidate for the theory of strong interactions, namely quantum chromodynamics, or color dynamics.

I want to introduce the bag model by invoking an analogy suggested by one of my graduate students, David Shalloway. Imagine a world in which the atmosphere consists not of air but of water. Energy could be expended to boil the water, creating bubbles of water vapor. Inside the bubbles the water-vapor molecules would behave, to a first approximation, like the weakly interacting particles of an ideal gas. In this hypothetical world water vapor could exist only inside the bubbles. A molecule of water vapor that returned to a liquid from a bubble would no longer be water vapor; it would cease to exhibit the properties of a particle of ideal gas.

The concept of hadronic matter in the bag model is similar. Hadrons are bubbles in space that contain quarks. Inside a bubble the quarks move about freely and independently, just as the water-vapor molecules move about inside a bubble. If an attempt is made to remove a quark from the bubble, the confining strong interaction between the quarks takes over. Like the water-vapor molecules, quarks can exist only in the bubbles enclosing them. In other words, hadrons act like ideal quark-vapor bubbles embedded in an atmosphere of ordinary space where quarks cannot exist.

Just as it takes a certain amount of latent heat to boil water, so it takes energy to "boil" a vacuum, that is, to convert energy into matter. Indeed, particle accelerators can be regarded as vacuum boilers. A large fraction of the kinetic energy of the accelerated particles is deposited on a small target. Most of the energy goes into making quarks and antiquarks that combine to form the hadrons found in the debris of the reac-

tion. In the bag model some of the energy also goes into the vacuum, pushing aside ordinary space in order to inflate the bubbles enclosing the quarks. The amount of energy needed to inflate a bubble is called the bag constant (B) and has a value of about 55 million electron volts per cubic fermi (one fermi being 10^{-15} meter).

If a particle is confined to a region of space that extends Δx in any direction, then according to the uncertainty principle of quantum physics the particle's momentum has a spread of about $\hbar/\Delta x$, where $\hbar$ is Planck's constant divided by 2π. This means that when quarks are confined to a bag of finite size, they move with a momentum on the order of $\hbar/\Delta x$ and hence exert a pressure on the inside surface of the bag. At some equilibrium value the pressure of the quarks on the inside surface balances the confining pressure B that keeps the matter inside the bag. It is this equilibrium value that determines the size of the hadrons. The pressure of ordinary space balancing the "uncertainty-principle pressure" of the quark vapor is somewhat analogous to the pressure of water balancing the pressure of the water vapor in a bubble.

It may look as though the problem of quark confinement has been trivially solved by simply postulating that the quarks are imprisoned in bags or bubbles. Yet a bag containing only a single quark would in effect be a quark, albeit a "fat" one. Such a particle would have the same charge and other properties of a single quark, and so the bag model as I have presented it so far does not account for the failure to observe an isolated quark. To account for quark confinement I must introduce the dynamical property of color. Each flavor of quark comes in three colors: "red," "blue" and "yellow." The antiquarks have anticolors: "antired," "antiblue" and "antiyellow." Quark color has nothing whatsoever to do with visual color. The word color is used because the way different colored quarks combine in quantum mechanics is reminiscent of the way visual colors combine. The best way to think about color (and also about flavor) is simply to imagine it as an extra variable associated with quarks. Just as a quark has a spin angular momentum of $1/2\hbar$ that can be oriented either up or down along any spatial axis, so quarks have flavors and colors that can also have different discrete values.

All the hadrons are "white," that is, they are colorless averages of the three colors. Roughly speaking, ordinary baryons consist of a red, a blue and a yellow quark, since these colors average to white, and ordinary mesons consist of a quark of one color and an antiquark of the corresponding anticolor. One way of establishing the truth of quark confinement is to show that the property of col-

SEPARATION OF CHARGES affects the force between them. When two opposite electric charges (*top*) are separated, the field lines bend more. The electric force gets weaker because only the component of the field lines that is lying along the direction between the charges has a net effect on the force. The force drops off as the square of the separation between the charges. When two quarks (color charges) are separated (*bottom*), the lines do not bend but continue to pull in the same direction. As a consequence of the fact that the color-field lines are parallel the color force between a quark and an antiquark is a constant.

or can never be observed, because to observe a color would be to see a free quark. It is now believed the strong interaction is based on quark color in much the same way that the electromagnetic interaction is based on electric charge. The other interactions (the electromagnetic, the weak and the gravitational) are all "color-blind."

The dynamical effects of quark color can best be elucidated by analogy to a much older concept: electric charge. Michael Faraday introduced the picture of a charge as a source of lines of electric force. A line of force is a curve in space that indicates the direction of the electric force a positive charge would be subjected to if it were placed on the curve. The number of lines in a given spatial region indicates the magnitude of the force; the greater the number of lines, the stronger the force. In Faraday's picture the lines of force begin on positive charges and end on negative charges because a positive charge is repelled by its own kind of charge and attracted by its opposite. The number of lines beginning or ending on a charge is a measure of the strength of the charge. Hence large charges generate strong electric forces.

Quark color can be thought of as a kind of charge. In electrodynamics there is only one type of charge (which can be either positive or negative), but in chromodynamics there are several types, corresponding roughly to the allowed color combinations. The color force acts not only to move the quarks but also to change their color, and so the color field itself must carry color. (An electric field acts only to move electric charges, not to change their sign or magnitude; this phenomenon was given mathematical generality in a theory of electrodynamics developed by C. N. Yang of the State University of New York at Stony Brook and Robert L. Mills of Ohio State University.) Quarks interacting in this way are consistent with the results of experiments where high-energy electrons were inelastically scattered by protons, the same experiments that revealed that protons are made up of quarks. (This consistency was pointed out by David J. Gross and Frank Wilczek of Princeton University and H. David Politzer of Cal Tech.) Fortunately it is not necessary to understand the mechanism of color transmutation in order to appreciate the relation between quark color and quark confinement.

Let me pursue the analogy between electric charge and color charge [see illustrations on pages 99 through 101]. It is helpful to think of a line of force that begins on one charge and ends on another as a kind of rubber band connecting the charges, because tension along the line of force is in part responsible for the attractive force between the charges. Adjacent lines exert a pressure on each other, and so the lines do not combine to form a single straight line linking the charges. The net force between the charges is the sum of the forces associated with each line. As the electric charges are separated the lines bend more. The force gets weaker because only the component of the field lines that lies along the direction between the charges has a net effect. The force drops off as the square of the separation between the charges. This inverse-squared relation is known as Coulomb's law.

One major difference between an ordinary electric field and the color field as described by the bag model is that the color lines of force connecting the quarks do not exist everywhere but are confined to the bag. At the surface of the bag where there are no quarks (and hence no color charges) the lines cannot begin or end, and since the lines cannot leave the bag, they must be tangent to its surface. At the surface the outward pressure of the color-field lines is not balanced by adjacent lines, since there are none, but by the confining bag pressure B. As a result the field lines are kept parallel to one another.

The fact that the lines are parallel intensifies the force acting between the charges because the lines all pull in the same direction. As the quarks (color charges) are separated the lines do not

		LEPTONS		QUARKS	
FIRST FAMILY	PARTICLE	e (ELECTRON)	ν_e (ELECTRON NEUTRINO)	u (UP)	d (DOWN)
	CHARGE	−1	0	2/3	−1/3
	MASS	$.5 \times 10^{-3}$	0? ($< .6 \times 10^{-7}$)	$\sim .5 \times 10^{-2}$	$\sim 1 \times 10^{-2}$
	POSITED	——	1929	1963	1963
	DISCOVERED	1898	1954	——	——
SECOND FAMILY	PARTICLE	μ (MU)	ν_μ (MUON NEUTRINO)	c (CHARMED)	s (STRANGE)
	CHARGE	−1	0	2/3	−1/3
	MASS	.11	0? ($< .57 \times 10^{-3}$)	~ 1.7	$\sim .3$
	POSITED	——	1957	1970	1963
	DISCOVERED	1936	1962	1974	1947
THIRD FAMILY	PARTICLE	τ (TAU)	ν_τ (TAU NEUTRINO)	t (TOP OR TRUTH)	b (BOTTOM OR BEAUTY)
	CHARGE	−1	0	2/3	−1/3
	MASS	1.78	0? ($< .25$)	?	~ 5
	POSITED	——	1975	1977	——
	DISCOVERED	1975	?	19??	1977
FOURTH FAMILY?		?	?	?	?
FIFTH FAMILY ?		?	?	?	?
. . . ?		?	?	?	?

FAMILIES OF ELEMENTARY PARTICLES each consist of two kinds of lepton (a particle that is not subject to the strong force and that shows no signs of an inner structure) and two "flavors," or kinds, of quark, each in three different colors. Like the leptons, the quarks seem to act like dimensionless points. The charge is measured in units of the charge of the proton, and the mass is given in terms of its energy equivalent in billions of electron volts (GeV). Although no quark has been seen in isolation, a discovery date is listed for each flavor that marks when a two-quark hadron consisting entirely of that flavor was found. The family-classification scheme is based on the fact that the quarks in a given family interact with each other in the same way that the quarks in any other family do. The leptons have been assigned to the families in order of increasing mass. Three families are known, and there is no reason blocking the discovery of still others.

bend but continue to pull in the same direction. The number of lines also remains the same. It is therefore a direct consequence of the fact the lines are parallel that the color force between a quark and an antiquark is a constant. The magnitude of the force is independent of the distance between the quarks, except when they are extremely close together.

The constant force between oppositely colored quarks is what keeps them together permanently. The magnitude of the force, which is the net tension in the color-field lines connecting the charges, is about 15 tons. (By way of contrast, the electric force of the proton on the electron in a hydrogen atom is about 10^{-11} ton.) The force of confinement is truly strong. Quarks are confined because the constancy of the force between a quark and an antiquark means that an arbitrarily large amount of energy is required to separate them. For example, to separate the quark and the antiquark in a meson by one inch would require the energy needed to make 10^{13} proton-antiproton pairs.

The only composites of quarks that have a finite mass are those in which the color-field lines all terminate inside a finite volume. Since the lines must begin and end on color charges, such composites must be colorless combinations of quarks. In other words, all hadrons are white, in agreement with observation. The problem of quark confinement has been solved by reducing it to color confinement. It is a consequence of the bag model that colors are confined. The way colors combine accounts for the fact that the baryon consists of only three quarks and the meson consists of only a single quark-antiquark pair. The simplest white composite consists of a quark and an antiquark of opposite colors. The next-simplest consists of a red, a blue and a yellow quark. It is impossible to make a colorless combination of quarks unless the number of quarks minus antiquarks is either zero or a multiple of three.

Chromodynamics in conjunction with the bag model can explain many aspects of hadron bahavior that once seemed paradoxical. For example, if a quark and an antiquark were spun, they would recede from each other like two skaters spinning in a circle holding the ends of an elastic rope, except that with quarks the rope consists of the color lines connecting the opposite color charges. When the system is spun, its angular momentum increases, the color lines stretch and energy is pumped into the color field the lines represent. As a result the mass of the system increases. The relation the model predicts between spin angular momentum and mass has been verified experimentally for all the hadrons [*see bottom illustration on this page*]. The experimental discovery of this rela-

	NAME	QUARK FLAVORS	SPINS	CHARGE	MASS (GEV)
MESONS	π^+	$u\bar{d}$	↑ ↓	+1	.14
	K^+	$u\bar{s}$	↑ ↓	+1	.50
	ρ^+	$u\bar{d}$	↑ ↑	+1	.76
	ϕ	$s\bar{s}$	↑ ↑	0	1.04
	J/ψ	$c\bar{c}$	↑ ↑	0	3.10
	D^{0*}	$c\bar{u}$	↑ ↑	0	2.00
	S^*	$s\bar{s}$ OR $(u\bar{u} + d\bar{d})$	↑ ↑ ↑ ↑	0	1.00
	γ	$b\bar{b}$	↑ ↑	0	9.46
BARYONS	p^+	uud	↑ ↑ ↑	+1	.938
	n	ddu	↑ ↑ ↑	0	.940
	Δ^{++}	uuu	↑ ↑ ↑	+2	1.2
	Σ^+	uus	↑ ↑ ↑	+1	1.4
	Ω^-	sss	↑ ↑ ↑	−1	1.65

QUARK CONTENTS of some prominent hadrons are listed here. The known quarks come in five flavors named "up" (*u*), "down" (*d*), "strange" (*s*), "charmed" (*c*) and "bottom" (*b*). The antiquarks are designated $\bar{u}$, $\bar{d}$, $\bar{s}$, $\bar{c}$ and $\bar{b}$. The hadrons have been divided into two classes depending on how they decay. The baryons are hadrons that decay ultimately into a proton; they are all made up of three quarks. The mesons are hadrons that decay entirely into photons and leptons or into proton-antiproton pairs; most mesons are made up of a single quark-antiquark pair. The arrows in the spin column indicate the direction in which the quarks are spinning.

tion was a main motivation for another model of quark confinement, the string model, in which two quarks cannot be separated because they are linked by a "string." The bag model provides a physical explanation of the nature of the string.

The derivation of the relation between mass and angular momentum on the basis of the bag model indicates why the same relation applies to mesons and baryons. The relation depends only on the color charge of the quarks at the ends of the particle. Since a baryon is

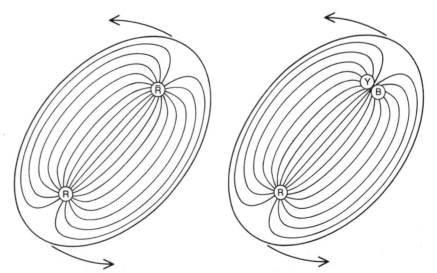

MESONS AND BARYONS have the same relation between angular momentum and mass, and the bag model is able to explain why that is the case. If the quark and the antiquark in a meson were spun (*left*), they would recede from each other like two skaters spinning in a circle holding the ends of an elastic rope, except that with quarks the rope consists of the color-field lines connecting the opposite color charges. When the system is spun, the angular momentum increases, the color lines stretch and energy is pumped into the color field the lines represent. As a result the mass of the system increases. The relation depends only on the difference in color between the quarks at each end of the particle. Since a baryon (*right*) is white, the two quarks (yellow and blue) at one end have a net color (antired) that is the opposite of the color (red) of the quark at the other end. Thus the quarks have the same color as the meson's antiquark.

white, the quarks at one of its ends have a combined color that is the opposite of the color of the quark at the other end. In other words, they must have the same net color as the antiquark in the meson.

The concept of color can also account for the production of hadrons. An electron and a positron (a positively charged electron) can be created as a pair, and so can a quark and an antiquark. Such a process requires a "polarizing field" acting to separate the color charges of a quark-antiquark pair in a meson. A new pair will appear spontaneously in the color field of the quark and antiquark of the meson. The color lines link the quark of the meson with the newly created antiquark and the antiquark of the meson with the newly created quark. Both quark-antiquark pairs are colorless, so that color lines do not join them. Because a constant force does not act between the pairs they can separate. In other words, the strong interaction can transform one meson into two others, for example a ρ meson into two pions.

The concept of color-field lines has also been successful in describing the self-annihilation of an electron and a positron at very high energies. If the particles carry a sufficiently large energy, they can generate an electromagnetic field strong enough to polarize the vacuum and create an oppositely charged quark-antiquark pair. If the pair consists of quarks of low mass, they fly apart just about as fast as the electron and the positron come together but not necessarily along the same axis. The quark and the antiquark are of course connected by color-field lines. As the quark and the antiquark rapidly move apart pairs of colored quarks that form mesons are created by the strong color field. The pressure of the color-field lines confines the diverging quark-antiquark pair to a tubular region of space about 10^{-15} meter thick. On the basis of the uncertainty principle the component of the pair's momentum that lies in the direction of the width of the tube is found to be the equivalent of only a few hundred million electron volts. Since the quark and the antiquark are rapidly diverging, the component of momentum along the axis defined by their motion is much higher. As a result the emerging hadrons also move primarily in this direction. Such hadron "jets" have been seen recently at SLAC and at other laboratories.

Electricity and magnetism are interrelated phenomena, as when electric charges in motion generate magnetic fields. Since color charge was introduced in analogy to electric charge, perhaps there is a kind of color magnetism. Indeed there is, and one of its consequences was one of the earliest successes of chromodynamics. This consequence was discovered independently by Howard Georgi, Alvaro De Rújula and Sheldon L. Glashow of Harvard University and by a group of us from M.I.T. It is a well-known empirical fact that currents flowing in the same direction through two parallel wires generate a magnetic force that pulls the wires together. By the same token currents flowing in opposite directions generate a magnetic force that pushes the wires apart.

Since quarks have spin angular momentum, they act as moving color charges and hence generate a color magnetic field. A spinning quark can be thought of as a color-current loop in which a fixed color charge circulates in a plane at right angles to the direction of the quark spin. Since the quarks in a meson are of opposite color, if the spins are parallel, then the color currents flow in opposite directions and so the quarks magnetically repel each other. On the other hand, if the spins are antiparallel, then the color currents flow in the same direction and so the color-magnetic force pulls the quarks toward each other. That makes particles whose quark spins are aligned more massive than those whose spins are opposed. This consequence of chromodynamics has been observed for all the hadrons. For

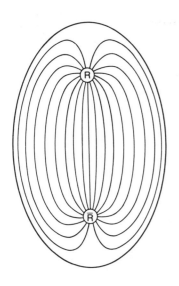
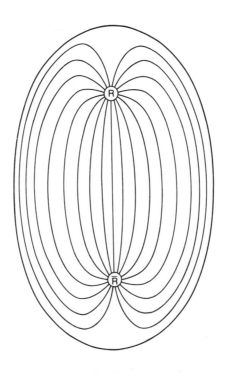
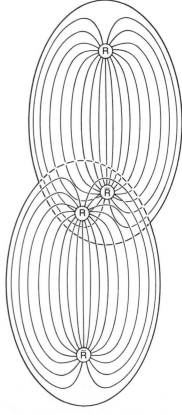

TRANSFORMATION OF A MESON (*left*) into two mesons requires a "polarizing field" acting to separate the color charges of the quark pair in the meson (*middle*). A new quark-antiquark pair appears spontaneously in the polarizing field between the quark and the antiquark of the meson (*right*). The color lines link the quark of the meson with the newly created antiquark and the antiquark of the meson with the newly created quark. Both quark-antiquark pairs are colorless, and so color lines do not join them. Because a constant force does not act between the pairs (as it does between the quark and the antiquark in each pair), they can separate and form two mesons.

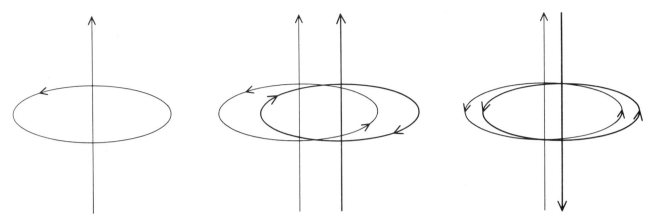

COLOR-MAGNETIC FIELD is generated by a moving quark (color charge) just as an ordinary magnetic field is generated by moving electric charges. A quark has spin angular momentum; it is therefore always spinning and can be treated as a color-current loop in which a fixed color charge circulates (counterclockwise for spin up) in a plane at right angles to the direction of the quark spin (*left*). Since the quarks in a meson are of opposite color, if the spins are parallel, then the color currents flow in opposite directions and so the quarks magnetically repel each other (*middle*). If the spins are antiparallel, then the color currents flow in the same direction and so the color-magnetic force pulls the quarks toward each other (*right*). That makes hadrons whose spins are aligned more massive than those whose spins are opposed.

example, the ρ meson whose quark spins are parallel is heavier than the π meson whose spins are antiparallel.

The fact that the atomic nucleus is made from a collection of neutrons and protons rather than directly from quarks is most likely a consequence of the magnitude of the color-magnetic forces. The energy of a conglomeration of up and down quarks is lower when they first cluster into colorless lumps (neutrons and protons), which in turn bind together (albeit relatively more weakly) to form an atomic nucleus. The mass density of the quarks in a hadron is only about twice as great as the mass density of the nucleons (protons or neutrons) in the atomic nucleus. This means that if nuclear matter could somehow be squeezed so that the mass density was increased by at least a factor of two, it could be energetically favorable for all the nucleons to lose their individual identity by merging to form a quark fluid. The transition from nuclear matter to quark matter is probably not a phase transition, such as the change from a gas to a liquid as the temperature is lowered, but a more gradual transition, such as the change from a monatomic gas to a diatomic gas as the temperature is lowered.

It would be extremely difficult to squeeze nuclear matter in the laboratory. It is believed, however, that nuclear matter is compressed by huge gravitational forces in a neutron star, the cold remnant of the explosion of a supernova. If the density of the squeezed neutron matter exceeds the density of the quarks in the neutron, the neutrons will merge to form a quark fluid. Hence the dense central region of a neutron star might actually consist of quark matter.

This possibility has important consequences for the stability of the neutron star. The matter of the star is squeezed by gravitational mass, mostly the mass of neutrons. Resistance to the squeeze is provided by the repulsive interaction of the neutrons and by the kinetic energy of the cold neutrons. It is thought the kinetic energy of the trapped quarks accounts for most of the neutron mass, and so the pressure would increase immensely if the neutrons merged and liberated the quarks. This effect would appreciably increase the allowed mass of a stable neutron star. Work is under way to discover the properties of quark matter so that the allowed mass can be calculated.

Since the color-field model describes so many features of the properties and interactions of hadrons, particle physicists have hoped that quark confinement would be a necessary consequence of chromodynamics. Is it possible the bag model of hadrons could follow from chromodynamics? The answer is yes. To understand how this might be so consider a possible analogy to superconductivity: the absence of electric resistance in a metal at a temperature near absolute zero. An applied magnetic force cannot penetrate a superconducting body. The magnetic field induces electron currents in a thin layer at the surface of the body; these currents in turn give rise to a magnetic field that is opposed to the applied field and is just large enough to cancel the applied field within the body. Under some conditions, however, the applied magnetic field can penetrate the body by creating in it a region of normal conductivity, in which lines of magnetic flux can be trapped. This phenomenon is called the Meissner effect.

The Meissner effect may have an analogue in chromodynamics. Perhaps the vacuum in quantum chromodynamics expels color-field lines just as a superconductor expels magnetic-field lines. In that case color fields might exist only where they expended energy to create a region of "normal vacuum" (analogous to the region of normal conductivity of the Meissner effect) in the surrounding vacuum. The color-expelling property of the surrounding vacuum would serve the same function as the bag pressure in the bag model, namely keeping the color-field lines confined to the inside of the hadrons. Charles G. Gross, Curtis G. Callan and R. Dashen of Princeton have made detailed calculations that support such a dual-vacuum model. Another model of this kind has been investigated by Holger B. Nielsen and M. Minomiya of the Niels Bohr Institute in Denmark. These models are at the forefront of continuing work on quark confinement.

Quantum chromodynamics and the bag model have been fairly successful in providing a mechanism for permanently sequestering quarks inside hadrons and hence for explaining many aspects of the strong force between quarks. Whether or not the mechanism succeeds ultimately depends on how well it works when it is exploited to make more refined and detailed calculations of the properties of hadrons. The phenomenon of quark confinement is perhaps the most paradoxical aspect of the quark model. A definitive explanation of the phenomenon will eliminate the last major detraction from the many successes of the quark hypothesis.

It is time to take stock of the picture of subatomic particles that has emerged from the quark model. Quarks are considered to be the elementary constituents of matter since they seem to interact like dimensionless points, but are they truly elementary? The history of science provides many examples of physical entities that were once thought to be elementary but that later turned out to be composite. Atoms, atomic nuclei, protons and neutrons were all thought to be indivisible at various times in this century until further experimentation unearthed their constituent parts. Is it possible that quarks will suffer the same fate? And what about the

ORBITAL ANGULAR MOMENTUM	PARALLEL SPINS			ANTIPARALLEL SPINS		
	PARTICLE	QUARK CONTENT	MASS (GEV)	PARTICLE	QUARK CONTENT	MASS (GEV)
0	Δ^+	uud	1.23	p	uud	.94
0	Σ^{+*}	uus	1.38	Σ^+	uus	1.19
0	Ξ^{-*}	dss	1.54	Ξ^-	dss	1.32
0	ρ^+	$u\bar{d}$	.77	π^+	$u\bar{d}$	.14
1	A_2^+	$u\bar{d}$	1.31	B^+	$u\bar{d}$	1.23
2	g^+	$u\bar{d}$	1.69	A_3	$u\bar{d}$	1.64
0	K^{-*}	$u\bar{s}$	.89	K^-	$u\bar{s}$	.49
1	K^{-**}	$u\bar{s}$	1.42	"Q"	$u\bar{s}$	1.3
2	K^{-***}	$u\bar{s}$	1.78	L^-	$u\bar{s}$	1.77

MASS EFFECTS OF COLOR MAGNETISM are shown here for various hadrons. The table clearly indicates that other things being equal a particle whose quark spins are parallel is more massive than one whose spins are antiparallel. The orbital angular momentum is given in units of Planck's constant divided by 2π. When the quarks rotate, they move away from each other as a result of the centrifugal force to which they are subjected in the rotating reference frame; the greater the rotation rate, the larger the separation. The color-magnetic interaction gets weaker with increasing separation. This means that as orbital angular momentum increases (and hence as rotation rate increases) the masses in the parallel and antiparallel states converge.

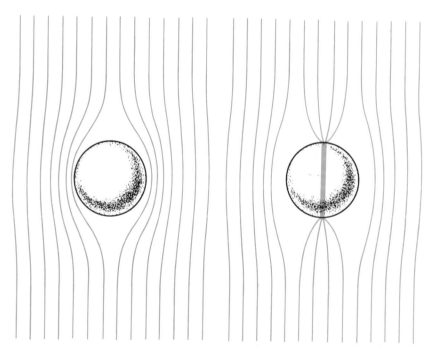

PROPERTIES OF A SUPERCONDUCTOR (a metal at temperatures near absolute zero that has no resistance to electric current) might serve as an analogue to how quark confinement could follow from chromodynamics, or color dynamics. An applied magnetic field cannot penetrate a superconducting body (*left*). The magnetic field induces electron currents in a thin layer at the surface of the body; the currents in turn give rise to a magnetic field that is opposite to the applied field and just large enough to cancel the applied field within the body. The only way the applied magnetic field can penetrate the body is for it to create a region of normal conductivity, in which lines of magnetic flux could be trapped (*right*). The vacuum in quantum chromodynamics might expel color-field lines just as a superconductor expels magnetic-field lines. In that case color fields might exist only where they expended energy to create in the surrounding vacuum a region of "normal vacuum" analogous to the region of normal conductivity. The color-expelling property of the surrounding vacuum would serve the same function as the bag in the bag model, namely keeping the color-field lines, and hence the quarks, confined.

leptons, which also are regarded as being elementary? Might they too have constituent parts?

The possibility that quarks or leptons have a substructure is all but ruled out by quantum mechanics and the theory of relativity. Quarks and leptons are thought to be no larger than 10^{-17} meter. For the electron this upper limit in size comes directly from experiments. On the basis of the uncertainty principle it is possible to calculate the momentum of a hypothetical constituent entity confined to a region of space 10^{-17} meter in diameter. The theory of relativity indicates that the kinetic energy of the constituent would be greater than its momentum multiplied by the speed of light. The kinetic energy of the constituent confined to a region 10^{-17} meter in diameter would increase the mass of the quark or lepton of which it was a part; the least energetic constituent would contribute a mass of at least 20 times the mass of the proton. Therefore it is extremely improbable that the electron, say, has constituents. For that to be the case there would have to be an accidental and nearly perfect cancellation of the binding energy of the constituent particles by their kinetic energy. In physics such accidents rarely happen, and so it is most unlikely that electrons (as well as other leptons and quarks) are composite in this sense.

If quarks and leptons are indeed elementary, nature cannot continue to complicate the picture of matter by creating new kinds of particles that are even more fundamental. Instead it seems to be complicating the picture by creating new flavors of quarks and new species of leptons. The known quarks and leptons can be classified by grouping them into families. Each family consists of four particles whose total electric charge is zero, namely two leptons (a particle and its associated neutrino) and two quark flavors (in the three different colors: red, blue and yellow). Three families have been identified so far. The first consists of the electron, the electron neutrino, the "up" quark and the "down" quark. The second consists of the muon, the muon neutrino, the "charmed" quark and the "strange" quark. The third consists of the tau particle, the tau neutrino, the "bottom" quark and an undetected but theoretically expected "top" quark.

The members of the first family play the largest role in naturally occurring phenomena. The other elementary particles, born chiefly in particle accelerators, have an ephemeral existence and are responsible only for extremely subtle effects in ordinary matter. They decay into members of the first family by the weak interaction.

This classification scheme is based on the fact that the four particles in a given family interact with one another in the same way that the four particles in any other family do. For example, within a family one quark flavor can be transformed into the other by the weak interaction. (That happens in ordinary radioactivity, where an up quark is changed into a down quark or vice versa.) By the same token one lepton can be transformed into the other. Quarks can also interact across family lines, although this interaction is considerably weaker. There are no known cases of leptons crossing family lines, of leptons changing into quarks or of quarks changing into leptons.

The members of the first family had all been identified by the late 1960's, when the up and down quarks were detected within the proton and the neutron. The particles in the second family were discovered between 1936 and 1974. The discovery in 1974 of the charmed quark, the last member of the second family, lent tremendous support to the quark model. Four years earlier the necessary existence of the charmed quark had been deduced from theoretical considerations by Glashow, John Iliopoulos and Luciano Maiani. The quark was subsequently discovered by Samuel C. C. Ting and his co-workers at the Brookhaven National Laboratory and by Burton D. Richter and his co-workers at SLAC. The third family came on the scene in 1975 with the discovery of the tau particle. Although it is not known with certainty that the tau's neutrino is distinct from the electron's neutrino, most particle physicists believe it is. Unless the top quark is extremely massive, it should be discovered within the next five years.

There is no theoretical reason blocking the discovery of still more families of particles. Some theorists believe insights into this possibility could come eventually from unified theories of weak and electromagnetic interactions. In the late 1960's one such theory was proposed independently by Steven Weinberg of Harvard and by Abdus Salam of the International Centre for Theoretical Physics in Trieste. Their proposal successfully accounts for many properties of the weak interaction. In its current form the proposal is fully compatible with the family-classification scheme, but it offers no explanation for the existence of the families. Although more far-reaching proposals that incorporate Weinberg's and Salam's theory provide such an explanation, they lack empirical support. Moreover, no compelling explanation has been offered for why the particles in a given family have the masses they do. Whereas a definitive model of quark confinement and a confirmation of the elementary nature of quarks and leptons would close a long chapter in particle physics, namely the search for the ultimate constituents of ordinary matter, the possible proliferation of families is opening another chapter.

III

FUTURE DIRECTIONS:
THE SEARCH FOR A
UNIFIED FIELD THEORY

9

Unified Theories of Elementary-Particle Interaction

by Steven Weinberg
July, 1974

*Physicists now invoke four distinct kinds of interaction,
or force, to describe physical phenomena. According to
a new theory, two, and perhaps three, of the forces are
seen to have an underlying identity*

One of man's enduring hopes has been to find a few simple general laws that would explain why nature, with all its seeming complexity and variety, is the way it is. At the present moment the closest we can come to a unified view of nature is a description in terms of elementary particles and their mutual interactions. All ordinary matter is composed of just those elementary particles that happen to possess both mass and (relative) stability: the electron, the proton and the neutron. To these must be added the particles of zero mass: the photon, or quantum of electromagnetic radiation, the neutrino, which plays an essential role in certain kinds of radioactivity, and the graviton, or quantum of gravitational radiation. (The graviton interacts too weakly with matter for it to have been observed yet, but there is no serious reason to doubt its existence.) A few additional short-lived particles can be found in cosmic rays, and with particle accelerators we can create a vast number of even shorter-lived species [*see top illustration on page 112*].

Although the various particles differ widely in mass, charge, lifetime and in other ways, they all share two attributes that qualify them as being "elementary." First, as far as we know, any two particles of the same species are, except for their position and state of motion, absolutely identical, whether they occupy the same atom or lie at opposite ends of the universe. Second, there is not now any successful theory that explains the elementary particles in terms of more elementary constituents, in the sense that the atomic nucleus is understood to be composed of protons and neutrons and the atom is understood to be composed of a nucleus and electrons. It is true that

the elementary particles behave in some respects as if they were composed of still more elementary constituents, named quarks, but in spite of strenuous efforts it has been impossible to break particles into quarks.

For all the bewildering variety of the elementary particles their interactions with one another appear to be confined to four broad categories [*see bottom illustration on page 112*]. The most familiar are gravitation and electromagnetism, which, because of their long range, are experienced in the everyday world. Gravity holds our feet on the ground and the planets in their orbits. Electromagnetic interactions of electrons and atomic nuclei are responsible for all the familiar chemical and physical properties of ordinary solids, liquids and gases. Next, both in range and familiarity, are the "strong" interactions, which hold protons and neutrons together in the atomic nucleus. The strong forces are limited in range to about 10^{13} centimeter and so are quite insignificant in ordinary life, or even on the scale (10^{-8} centimeter) of the atom. Least familiar are the "weak" interactions. They are of such short range (less than 10^{-15} centimeter) and are so weak that they do not seem to play a role in holding anything together. Rather, they are manifested only in certain kinds of collisions or decay processes that, for whatever reason, cannot be mediated by the strong, electromagnetic or gravitational interactions. The weak interactions are not, however, irrelevant to human affairs. They provide the first step in the chain of thermonuclear reactions in the sun, a step in which two protons fuse to form a deuterium nucleus, a positron and a neutrino.

From this brief outline one can see

that a certain measure of unification has been achieved in making sense of the world. We are still faced, however, with the enormous problem of accounting for the baffling variety of elementary-particle types and interactions. Our prospects for further progress would be truly discouraging were it not for the guidance we receive from two great products of 20th-century physics: the development of quantum field theory and the recognition of the fundamental role of symmetry principles.

The Necessity of Fields

Quantum field theory was born in the late 1920's through the union of special relativity and quantum mechanics. It is easy to see how relativity leads naturally to the field concept. If I suddenly give one particle a push, this cannot produce any instantaneous change in the forces (gravitational, electromagnetic, strong or weak) acting on a neighboring particle because according to relativity no signal can travel faster than the finite speed of light. In order to maintain the conservation of energy and momentum at every instant, we say that the pushed particle produces a field, which carries energy and momentum through surrounding space and eventually hands some of it over to the neighboring particle. When quantum mechanics is applied to the field, we find that the energy and momentum must come in discrete chunks, or quanta, which we identify with the elementary particles. Thus relativity and quantum mechanics lead us naturally to a mathematical formalism, quantum field theory, in which elementary-particle interactions are explained by the exchange of elementary particles themselves.

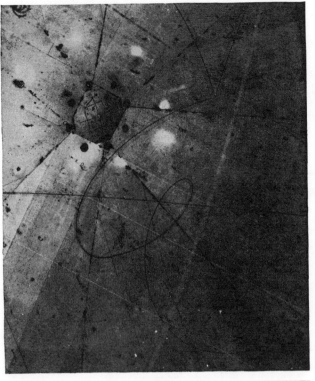

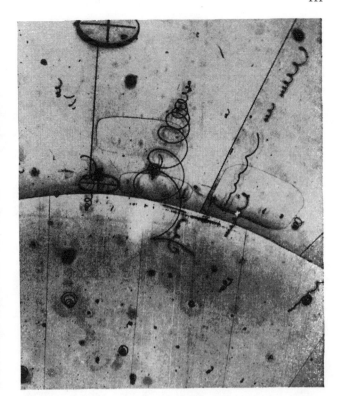

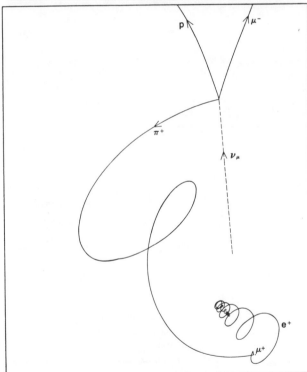

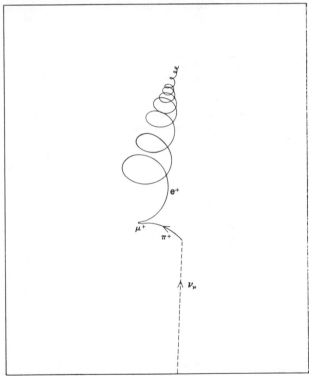

EVIDENCE FOR NEUTRAL CURRENTS, the existence of which would support theories showing a connection between electromagnetic interactions and weak interactions, was recently obtained in an experiment conducted at the Argonne National Laboratory with a neutrino beam from the zero-gradient synchrotron and with a 12-foot bubble chamber filled with liquid hydrogen. The bubble-chamber photograph at left and the map below it show an example of a familiar kind of charged-current process ($\nu_\mu + \mathrm{p} \rightarrow \mu^- + \mathrm{p} + \pi^+$) in which a unit of electric charge is exchanged between leptons (ν_μ, μ^-) and other particles. The photograph at the right and the map below it show an example of a neutral-current process ($\nu_\mu + \mathrm{p} \rightarrow \nu_\mu + \mathrm{n} + \pi^+$) distinguished by the absence of outgoing negative muon (μ^-) or proton (p) tracks. In such photo-graphs tracks are left only by charged particles, so that the incoming neutrino (ν_μ) and the outgoing neutrino and neutron (n) in the neutral-current process are invisible. Moreover, the bubble chamber is subjected to an intense magnetic field, which causes charged particles to follow curved tracks, clockwise for negative charge and counterclockwise for positive charge. In both of these photographs the positive pion (π^+) is seen to decay into a positive muon (μ^+), which then decays into a positive electron (e^+), visible as a tightly wound spiral. This experiment is more recent than similar ones that have been conducted at European Organization for Nuclear Research (CERN) and at National Accelerator Laboratory. It provides the first evidence for the specific neutral-current reactions $\nu_\mu + \mathrm{p} \rightarrow \nu_\mu + \mathrm{n} + \pi^+$ and $\nu_\mu + \mathrm{p} \rightarrow \nu_\mu + \mathrm{p} + \pi^0$.

	PARTICLE	SYMBOL	CHARGE	MASS (10⁶ ELECTRON VOLTS)	LIFETIME (SECONDS)
	PHOTON	γ	0	0	∞
LEPTONS — NEUTRINO	NEUTRINO	$\nu_e \; \bar{\nu}_e$	0	0	∞
	NEUTRINO	$\nu_\mu \; \bar{\nu}_\mu$	0	0	∞
	ELECTRON	$e^\pm$	$\pm e$	0.511	∞
	MUON	$\mu^\pm$	$\pm e$	105.66	2.199×10^{-6}
MESONS — PION	PION	$\pi^\pm$	$\pm e$	139.57	2.602×10^{-8}
	PION	π°	0	134.97	0.84×10^{-16}
	KAON	$K^\pm$	$\pm e$	493.71	1.237×10^{-8}
	KAON	K°	0	497.71	0.882×10^{-10}
	ETA	η	0	548.8	2.50×10^{-17}
BARYONS — PROTON	PROTON	$p \; \bar{p}$	$\pm e$	938.259	∞
	NEUTRON	$n \; \bar{n}$	0	939.553	918
	LAMBDA HYPERON	$\Lambda \; \bar{\Lambda}$	0	1,115.59	2.521×10^{-10}
	SIGMA HYPERON	$\Sigma^+ \; \bar{\Sigma}^+$	$\pm e$	1,189.42	8.00×10^{-11}
	SIGMA HYPERON	$\Sigma^\circ \; \bar{\Sigma}^\circ$	0	1,192.48	$< 10^{-14}$
	SIGMA HYPERON	$\Sigma^- \; \bar{\Sigma}^-$	$\pm e$	1,197.34	1.484×10^{-10}
	CASCADE HYPERON	$\Xi^\circ \; \bar{\Xi}^\circ$	0	1,314.7	2.98×10^{-10}
	CASCADE HYPERON	$\Xi^- \; \bar{\Xi}^-$	$\pm e$	1,321.3	1.672×10^{-10}
	OMEGA HYPERON	$\Omega^- \; \bar{\Omega}^-$	$\pm e$	1,672	1.3×10^{-10}

PARTIAL LIST OF OBSERVED ELEMENTARY PARTICLES identifies all those with lifetimes greater than 10^{-20} second. Apart from the photon, all the observed particles fall into two broad families: leptons and hadrons. The leptons are either massless or low-mass particles that do not take part in the "strong" interactions; the hadrons are heavier and do take part. Hadrons are further divided into mesons and baryons according to rotational angular momentum and other properties. A symbol with a bar above it denotes an antiparticle. Neutrinos and antineutrinos are of two types: electron-type, ν_e, and muon-type, ν_μ. In three cases (photon, neutral pion and eta meson) the particle is its own antiparticle. Charges are given in units of charge, e, on the electron, equal to 1.602×10^{-19} coulomb. Masses are in energy units; one million electron volts (MeV) equals 1.783×10^{-27} gram.

	GRAVITATIONAL	ELECTRO-MAGNETIC	STRONG	WEAK
RANGE	∞	∞	$10^{-13} - 10^{-14}$ CM.	$<< 10^{-14}$ CM.
EXAMPLES	ASTRONOMICAL FORCES	ATOMIC FORCES	NUCLEAR FORCES	NUCLEAR BETA DECAY
STRENGTH (NATURAL UNITS)	$G_{NEWTON} = 5.9 \times 10^{-39}$	$e^2 = \frac{1}{137}$	$g^2 \approx 1$	$G_{FERMI} = 1.02 \times 10^{-5}$
PARTICLES ACTED UPON	EVERYTHING	CHARGED PARTICLES	HADRONS	HADRONS LEPTONS
PARTICLES EXCHANGED	GRAVITONS	PHOTONS	HADRONS	?

FOUR TYPES OF INTERACTION among particles are believed to account for all physical phenomena. "Range" is the distance beyond which the interaction effectively ceases to operate. In two cases the range is believed to be infinite. "Strength" is a dimensionless number that characterizes the strength of the force under conditions typical of current observations. Thus the gravitational force is some 39 orders of magnitude weaker than the strong force.

It is a simple consequence of the uncertainty principle in quantum mechanics (which states that the uncertainties in our knowledge of the momentum and the position of a particle are inversely proportional to each other) that the range of the force should be inversely proportional to the mass of the exchanged particle. (For an exchanged mass equal to that of the proton the range is about 2×10^{-14} centimeter.) Thus electromagnetism and gravitation, which seem to be of infinite range, are due to the exchange of particles of zero mass: the familiar photon and the hypothetical graviton. The strong interactions are generally believed to arise from the exchange of a large variety of strongly interacting particles, including protons, neutrons, mesons and hyperons of various kinds. Since the weak interactions have a much shorter range than the strong interactions, they must be produced by the exchange of much heavier particles, presumably particles too heavy to have yet been created with existing accelerators.

The Intermediate Vector Boson

For many years it has been speculated that there may be a deep relation between the weak interactions and the electromagnetic interactions, with the difference in their apparent strengths being due simply to the large mass of the particle exchanged in the weak interactions. This hypothesis is supported by the observation that the angular momentum exchanged in weak processes such as nuclear beta decay [see illustration on opposite page] has the same value as the angular momentum of a single photon (equal to the Planck constant: 1.0546×10^{-27} erg-second). In fact, the hypothetical particle, or quantum, exchanged in weak interactions has long had a name: the intermediate vector boson. (The term "vector" is used because any particle with this angular momentum is usually described by a field that is a four-dimensional vector, like the vector potential used to describe the photon in James Clerk Maxwell's theory of electromagnetism. The term boson refers to the entire class of particles whose angular momentum is an integer multiple of Planck's constant.)

If we assume that the intrinsic interaction strength of the putative intermediate vector boson that is exchanged in the weak interactions is the same as it is for a photon, then the weak force will have the same strength as the electromagnetic force at short distances; it only appears weaker because of its much

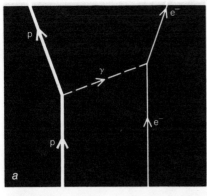
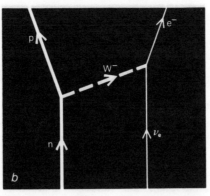
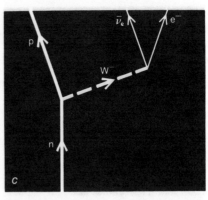

ELECTROMAGNETIC AND WEAK PROCESSES exhibit striking similarities when depicted in the form of Feynman diagrams. Such diagrams symbolize the interactions that underlie subnuclear phenomena, for example the collision between two particles, which physicists refer to as a scattering event. Thus diagram *a* indicates that the electromagnetic scattering of an electron by a proton is due to the exchange of a photon, which transfers energy and momentum from one particle to another. Since time proceeds upward in these diagrams, the photon in *a* is traveling from the proton to the electron, but the diagram is intended also to cover the equally important case in which the photon is traveling in the opposite direction on a trajectory from lower right to upper left. It is precisely this feature of lumping together different processes in a single graph that constitutes the great conceptual value of the "language" of the Feynman diagrams. The scattering of a neutrino by a neutron (*b*) represents a weak interaction in which a heavy particle as yet undetected, the intermediate vector boson, $W^{\pm}$, is believed to play a role analogous to that of the photon in electromagnetic scattering. Here the W particle is assigned a negative charge because it is assumed to be traveling from left to right. It could equally well be regarded as carrying a positive charge and traveling from right to left. The intermediate vector boson is also thought to mediate the radioactive decay of a neutron into a proton, an electron and an antineutrino (*c*). Note that diagram *b* can be obtained from diagram *a* simply by changing some of the particles into others of different charge. Diagram *c* can be obtained from diagram *b* by replacing the incoming neutrino by an outgoing antineutrino. Note also that the total charge is conserved at each vertex in the diagrams.

shorter range. The effect of the weak force is reduced in any given process by a factor given by the square of the ratio of the typical masses involved in the process to the mass of the intermediate vector boson. For processes characterized by masses comparable to the mass of the proton, the weak force is roughly 1,000 times weaker than the electromagnetic force. Hence, taking the square root, we conclude that the mass of the intermediate vector boson is roughly 30 proton masses. By applying the conservation of charge to processes such as nuclear beta decay we also see that the intermediate vector boson must carry either a positive or a negative electric charge equal in magnitude to the charge of the proton or of the electron.

A quantum field theory tells us how to calculate the rate for any process in terms of a sum of individual processes, each symbolized by a Feynman diagram such as the one shown above and elsewhere in this article. This useful method for visualizing subnuclear events was introduced some 25 years ago by Richard P. Feynman, and led to a solution of the one great problem that had plagued quantum field theory almost from its birth: the problem of infinities. It was found in the 1930's that the contributions produced by processes more complicated than single-particle exchanges usually turn out to be infinitely large. In fact, the electrostatic repulsion within a single electron produces an infinite self-energy, which manifests itself whenever a photon is emitted and reabsorbed by the same electron [*see illustration on next page*]. These infinities arise only in Feynman diagrams with loops, and they can be traced to the infinite number of ways that energy and momentum can flow through the loop from one particle to another. As is usually the case when paradoxes arise in science, the problem of infinities is both a curse and a blessing: a curse because it keeps us from getting on with calculations we would like to perform, and a blessing because when the solution is found, it may work only for a limited class of theories, among which one hopes to find the true theory.

That is just what seems to have happened with the problem of infinities. In the late 1940's a group of young theoreticians working independently (Feynman, then at Cornell University, Julian Schwinger at Harvard University, Freeman J. Dyson at the Institute for Advanced Study and Sin-itiro Tomonaga in Japan) found that in a certain limited class of field theories the infinities occur only as "renormalizations," or corrections, of the fundamental parameters of the theory (such as masses and charges) and can therefore be eliminated if one identifies the renormalized parameters with the measured values listed in tables of the fundamental constants. For example, the measured mass of the electron is the sum of its "bare" mass and the mass associated with its electromagnetic self-energy. In order for the measured mass to be finite, the bare mass must have a negative infinity that cancels the positive infinity in the self-energy. One simple version of the field theory of electromagnetic interactions not only was found to be renormalizable in the sense that all infinities could be eliminated by a renormalization of the electron's mass and charge but also led to electrodynamic calculations whose agreement with experiment is without precedent in physical science. Thus the theory predicts that the value of the magnetic moment of the electron (in natural units) is 1.0011596553, whereas the observed value is 1.0011596577. The uncertainty in both figures is in the ninth place: ±.0000000030.

Tougher Infinity Problems

In spite of this stunning success, attempts to construct renormalizable field theories of the other elementary-particle interactions long proved to be unsuccessful. For the strong interactions there was no lack of possible renormalizable theories; rather, the trouble was (and indeed still is) that the strength of the interaction invalidates any simple approximation scheme that might be used to draw consequences from a given field theory that could be checked by experiment. (Roughly speaking, the probability of exchanging a set of strongly interacting particles in a high-energy collision is independent of the number of particles exchanged, so that very com-

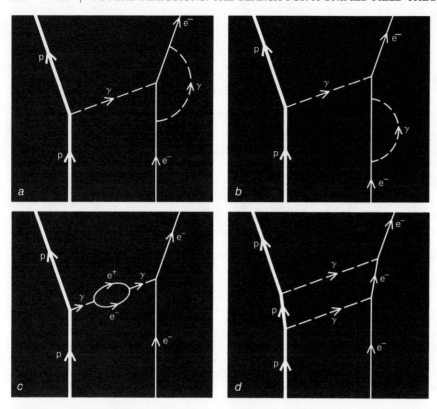

HIGHER-ORDER CONTRIBUTIONS TO SCATTERING RATES were formerly impossible to calculate when a photon was emitted and reabsorbed by the same electron (*a, b*) or when a photon gave rise to an electron-positron pair that subsequently recombined into a photon (*c*). When physicists see "loops" of this kind in Feynman diagrams, they are prepared to encounter infinities in trying to calculate reaction rates. Thus in diagram *b* the electrostatic repulsion within a single electron manifests itself as an infinite self-energy. It was found in the late 1940's that such infinities can be handled by redefining the mass and charge of the electron, a process called renormalization. Infinity problem does not arise, however, in scattering events such as that in *d*, where the loop has four corners or more.

plicated exchanges have to be taken into account in even the lowest approximation.)

For the gravitational interactions we have a well-known field theory, Einstein's general theory of relativity, which accounts very well for phenomena on the scale of the solar system but seems not to be renormalizable and presumably therefore needs modification for phenomena at very short distances. The problem here is the opposite of that for the strong interactions: gravitational effects are so weak that one can get no help from experimental measurements, at their current level of precision, in finding the correct theory.

The weak interactions present an intermediate case: they are strong enough so that good experimental data are available (although nowhere near as copious as for the strong interactions) and yet weak enough so that approximate calculations are practicable. Even though the weak interactions are believed to be similar to the electromagnetic interactions, however, the theory, as it existed until a few years ago, does not appear

to be renormalizable. To be more specific, the exchange of pairs of intermediate vector bosons in processes such as neutron-neutrino scattering [*see illustration on page 117*] leads to infinities that cannot be absorbed in a renormalization of the parameters of the theory. Hence although the quantum field theory of intermediate vector bosons gave a perfectly good approximate picture of the observed weak interactions, it broke down as soon as it was pushed beyond the lowest approximation.

What is the difference between photons and intermediate vector bosons that makes the infinities so much worse for the latter? A detailed analysis enables us to trace the difference back to the fact that the photon has zero mass whereas the intermediate vector boson has ponderable mass. Like all other zero-mass particles, the photon can exist as a superposition of at most two pure states, characterized by left or right circular polarization, in which the axis of rotation is respectively in the same direction as the direction of motion or in the opposite direction. On the other hand, the inter-

mediate vector boson, like any other massy particle whose angular momentum equals Planck's constant, can exist in any one of three states, characterized by an axis of rotation that points in the direction of motion, points in the opposite direction or points in a direction perpendicular to the direction of motion. It is the exchange of intermediate vector bosons whose axes of rotation are perpendicular to the direction of motion that produces the nonrenormalizable infinities.

Types of Symmetry

Before we can see how this problem can be resolved we must first consider the role that symmetry principles have come to play in theoretical physics. Considerations of symmetry have always been important in science, but they acquired special significance with the advent of quantum mechanics. That is because the energy or mass levels of any quantum-mechanical system subject to a symmetry principle are generally required to form certain well-defined and easily recognized families. (In mathematical language one says that the collection of all the mathematical operations on fields that leave the form of the field equations unchanged constitute a "group"; the levels with a given energy or mass are said to form a representation of that group.)

For example, the quantum-mechanical equations that describe the hydrogen atom obey the symmetry principle that all spatial directions are equivalent. As a result the energy levels of hydrogen form families with an odd number of members (1, 3, 5 and so on), the levels within each family being distinguished by the orientation of the axis of rotation [*see top illustration on opposite page*]. When we look at a table showing the masses of elementary particles, we find a similar grouping into families: the proton and the neutron have nearly equal mass; the three sigma hyperons (Σ^+, Σ^0, Σ^-) likewise have nearly the same mass, and so on. For this reason it is believed that the field equations of elementary-particle physics obey isotopic spin symmetry, a symmetry principle analogous to rotational symmetry except that the "rotation" alters the value of the particle's electric charge rather than its spatial orientation. In the early 1960's it was further realized that the various particle pairs, triplets and so on are themselves grouped into larger superfamilies of eight, 10 or even more members, reflecting an approximate symmetry larger than isotopic spin symmetry [*see bottom illustration at right*].

All these symmetry principles require that the field equations do not change when we simultaneously perform well-defined "rotations" on some labeled characteristic of a family or superfamily of particles everywhere in space. One can imagine a much more powerful requirement: that the equations should not change when we perform such rotations of labeled characteristics independently at each point in space and time. The first and less stringent kind of symmetry operation is comparable to giving each apple in a basket the same rotation from one orientation in space to another. The second and more general kind of symmetry operation is comparable to rotating each apple in the basket separately to different new orientations. Invariance under the second kind of symmetry operation is known as a gauge symmetry.

It has been known for many years that the Maxwell field equations of electromagnetism obey a gauge symmetry, based on the group of rotations in two rather than in three directions. Indeed, the logic can be turned backward: assuming this gauge-symmetry principle, one can deduce all the properties of electromagnetism, including Maxwell's equations and the fact that the mass of the photon is zero. It is difficult to conceive of any better example of the power of symmetry principles in physics.

Appearance and Reality

Our hopes of perceiving an underlying identity in the weak and the electromagnetic interactions lead us naturally to suppose there may be some larger gauge symmetry that forces the photon and the intermediate vector boson into a single family. (Indeed, the mathematical theory of generalized gauge symmetries has been understood since the work in 1954 of C. N. Yang and Robert L. Mills, who were then working at Brookhaven National Laboratory.) For this to be possible, however, the intermediate vector boson, like the photon, would have to have zero mass, and we have already seen that its mass is actually much greater than that of any known particle. How can there be any family connection between two such different particles?

The answer to this conundrum lies in considerations of appearance and reality. Inasmuch as symmetry principles govern the form of the field equations, they are generally regarded as providing information about the laws of nature on the deepest possible level. Is it conceivable for a symmetry principle to be valid on this level and yet not be manifest in

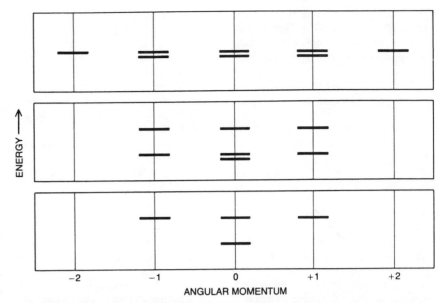

SYMMETRY PRINCIPLES require the energy levels of an atom, such as the lower levels of hydrogen shown here, to cluster in well-defined families. Each quantum state of the hydrogen atom is indicated by a short bar. The value of the energy is indicated schematically by the bar's vertical position. (The breaks between panels indicate energy gaps.) The value of the angular momentum around any fixed direction (in units of Planck's constant) is indicated by the horizontal position of the bar. The exact equality of energies within the various triplets, quintuplets and so on is a consequence of the rotational symmetry of the equations that describe the atom, whereas the approximate equality of energies within the various larger families is dictated by more detailed features of the dynamics, such as the weakness of the magnetic coupling between proton and electron, the small value of the electron's charge and the decrease in electrostatic attraction with distance squared.

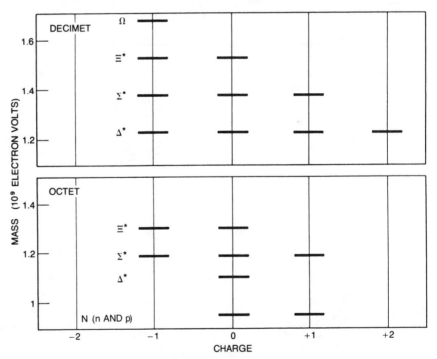

FAMILIES OF ELEMENTARY PARTICLES are believed to be a consequence of a symmetry principle known as isotopic spin symmetry, analogous to the rotational symmetry that produces the families of quantum states within the hydrogen atom. The grouping of these families of elementary particles into superfamilies (octets, decimets and so on) was proposed independently in the early 1960's by Murray Gell-Mann and Yuval Ne'eman. The bars' vertical position indicates the mass of the particles. Their horizontal position indicates the electric charge in units of the proton's charge. Masses differ slightly from particle to particle even within families, but these differences are too small to be shown here. Particles with an asterisk are very-short-lived states, not included in the table at top of page 112.

the masses and other observed properties of physical particles? The familiar phenomenon of ferromagnetism provides an example of how this can happen.

The equations governing the electrons and iron nuclei in a bar of iron obey rotational symmetry, so that the free energy of the bar is the same whether one end is made the north pole by magnetization or the south. At high temperatures the curve of energy versus magnetization has a simple U shape that has the same rotational symmetry as the underlying equations [see illustration below]. The equilibrium state, the state of lowest energy at the bottom of the U, is also a state of zero magnetization, which shares this symmetry. On the other hand, when the temperature is lowered, the lowest point on the U-shaped curve humps upward so that the curve resembles a W with rounded corners. The curve still has the same rotational symmetry as the underlying equations, but now the equilibrium state has a definite nonzero magnetization, which can be either north or south but which in either case no longer exhibits the rotational symmetry of the equations. We say in such cases that the symmetry is spontaneously broken. A tiny physicist living inside the magnet might not even know that the equations of the system have an underlying rotational symmetry, although we, with our superior perspective, find this easy to recognize. Reasoning by analogy, we see that a symmetry principle might thus be exactly true in a fundamental sense and yet not be visible at all in a table of elementary-particle masses.

The first proposed example of a broken symmetry of this kind in elementary-particle physics was a non gauge symmetry known as chiral symmetry. (The term chiral is from the Greek for hand, and it is used here because the symmetry consists of independent three-dimensional rotations on fields of left-handed or right-handed polarization. This symmetry contains within it the unbroken three-dimensional rotation group of isotopic spin symmetry.) Chiral symmetry has led to great successes in predicting the properties of low-energy pi mesons, but a discussion of such matters would take us too far afield.

New Theories and Predictions

In 1967 I suggested that the weak and electromagnetic interactions are governed by a broken gauge-symmetry group. (A similar suggestion was made independently some months later by Abdus Salam of the International Center for Theoretical Physics in Trieste.) The proposed group contains within it the unbroken gauge-symmetry group of electromagnetism and therefore requires the photon to have zero mass, but the other members of the photon's family are associated with broken symmetries and therefore pick up a large mass from the symmetry-breaking. In the simplest version of this theory the relatives of the photon would consist of a charged intermediate vector boson (long referred to as the W particle) with a mass greater than 39.8 proton masses plus an additional neutral intermediate vector boson (which I called the Z particle) with a

mass greater than 79.6 proton masses. (A theory of this kind is more closely analogous to superconductivity than to ferromagnetism: in a superconductor electromagnetic gauge symmetry is broken and the photon itself acquires mass, as is shown by the fact that a magnetic field can penetrate only a short distance into a superconductor. In particle physics the appearance of vector boson masses in this way is called the Higgs mechanism because it first became known as a mathematical possibility through a 1964 paper by Peter Higgs of the University of Edinburgh.)

At the time I proposed my theory there was no experimental evidence for or against it and no immediate prospect of getting any. There was, however, an internal test of the theory that could be made without help from experiment. We have seen that the infinities in the quantum field theory of pure electromagnetism can be renormalized away, whereas this cannot be done with the existing theory of weak interactions in which intermediate vector bosons have mass. Thus one can ask: Does a field theory become renormalizable when the intermediate vector bosons belong to the same family as the photon and acquire mass only through the spontaneous breakdown of a gauge symmetry? I had suggested in 1967 that this might be the case, but the renormalizability of the theory was not demonstrated until four years later, when it was first shown by Gerhard 't Hooft, then a graduate student at the University of Utrecht. (The proof has since been made more rigorous through the work of many theorists, par-

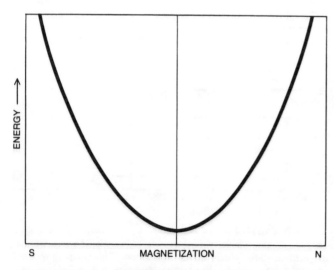
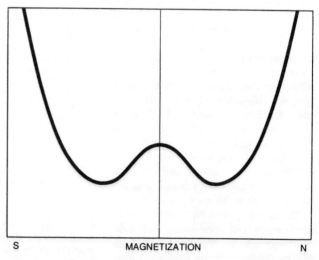
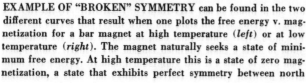

EXAMPLE OF "BROKEN" SYMMETRY can be found in the two different curves that result when one plots the free energy v. magnetization for a bar magnet at high temperature (left) or at low temperature (right). The magnet naturally seeks a state of minimum free energy. At high temperature this is a state of zero magnetization, a state that exhibits perfect symmetry between north and south. At low temperature the equilibrium state shifts to one of nonzero magnetization, which can be either north or south, even though the free-energy curve is still perfectly symmetrical between north and south. In this case physicists say that the symmetry is spontaneously broken. The author invokes a similar breaking of symmetry to unify the electromagnetic and weak interactions.

ticularly B. W. Lee, J. Zinn-Justin, M. Veltman and 't Hooft himself.) It turns out that the various multiparticle exchanges involving photons, charged intermediate vector bosons, neutral intermediate vector bosons and other particles add up so as to cancel all nonrenormalizable infinities [*see top illustration on next two pages*].

Once the renormalizability of the theory was established it became clear that the long-sought goal of a unified field theory of weak and electromagnetic interactions might finally be at hand. It then became crucially important to test the theory against experiment. Until such time as intermediate vector bosons can be produced directly the best way to test the theory is to look for effects attributable to the newly predicted neutral intermediate vector boson —the new Z particle—that must appear in the same family as the photon and the charged intermediate vector bosons. The neutral boson does not contribute to processes such as beta decay, in which a charge must be exchanged between the nucleus and the emitted particles. It does, however, contribute along with the charged bosons in processes such as the scattering of "ordinary" neutrinos by electrons [*see "a" and "b" in bottom illustration on next two pages*] and materially changes their rate. Finally, there are processes such as the elastic scattering of "muon-type" neutrinos by electrons and the elastic scattering of any type of neutrino by protons or neutrons [*see "c" and "d" in bottom illustration on next two pages*] that could be produced only by exchanges of neutral intermediate vector bosons.

For some years these neutral current processes, as they are called, remained at the edge of detectability, and many physicists doubted their existence. Within the past year, however, evidence for neutral-current processes has at last begun to appear. A pan-European collaboration involving some 55 investigators from seven different institutions, working at the European Organization for Nuclear Research (CERN) in Geneva, has found two events in which muon-type antineutrinos are scattered by electrons and several hundred events in which they are scattered by protons or neutrons. Such scattering events can apparently be explained only by the exchange of a neutral intermediate vector boson, or Z particle, and are therefore direct evidence for a new kind of weak interaction. Moreover, the inferred collision rates agree well with rates predicted by the new theory. An American consortium working at the National

Accelerator Laboratory in Batavia, Ill., and another group working at the Argonne National Laboratory have apparently also found neutral-current events [*see illustration on page 111*]. Further experiments aimed at the detection of neutral-current processes and the measurement of their rates are in train at various laboratories in Western Europe, the U.S. and the U.S.S.R.

The existence of neutral-current processes is not yet definitely established, and in any case the general idea of a renormalizable unified field theory based on a spontaneously broken gauge symmetry does not depend absolutely on the existence of neutral-current processes. For instance, in one model suggested by Howard Georgi and Sheldon L. Glashow of Harvard the photon and the charged intermediate vector boson form a family by themselves, although this simplification is achieved at the cost of introducing new particles of other kinds. (There are now a host of other ingenious models suggested by more theorists than can be named here.) There is no doubt, however, that the apparent detection of neutral-current processes has brought welcome encouragement to field theorists.

Some Further Implications

At the same time that experimentalists have been working to test the consequences of unified weak and electromagnetic field theories, theoreticians have been discovering that the new theories freshly illuminate a number of outstanding problems. One, for instance, has to do with the dynamics of the giant stellar explosions known as supernovas. It is believed that supernovas occur at a certain point in the life of a very massive star when the core of the star becomes unstable and begins to implode, or collapse. It has long been a puzzle how the implosion can be reversed and become an explosion, and how the star can shed enough of its outer layers to reach stability as an ultradense neutron star, only 10 or 20 kilometers in diameter. (There is observational evidence that at least two pulsars, believed to be rapidly rotating neutron stars, are embedded in the remnants of past supernovas.)

In 1966 Stirling A. Colgate and R. H. White of the New Mexico Institute of Mining and Technology suggested that the outer layers of an exploding star might be blown off by the pressure of neutrinos produced in the hot stellar core, but detailed calculations by James Wilson of the Lawrence Livermore Laboratory of the University of California,

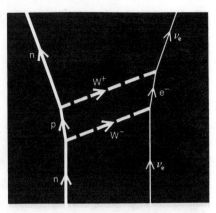

INFINITY PROBLEM ARISES in calculating the rates of neutron-neutrino scattering events involving the sequential (and hypothetical) exchange of two intermediate vector bosons, W^- and W^+. The first exchange converts the neutron into a proton and the neutrino into an electron. The second exchange restores the original cast of characters. Note that the Feynman diagram of this neutron-neutrino event has the same form as diagram *d* in illustration on page 54, in which infinity problem does not arise. The author's recent work, however, indicates how infinities can be removed in processes involving the intermediate vector boson.

using the then available theories of weak interactions, did not support the conjecture. It has recently been pointed out by Daniel Freedman of the State University of New York at Stony Brook that neutral currents can produce "coherent" neutrino interactions, in which a neutrino interacts with an entire nucleus rather than with its individual neutrons and protons. This leads to a much stronger interaction between neutrinos and the relatively heavy nuclei, chiefly the nuclei of iron, in the outer layers of the stellar core. According to Wilson's latest calculations, the increased neutrino pressure is apparently sufficient to produce a supernova.

Another old problem that may be solved through the development of unified gauge theories of weak and electromagnetic interactions concerns the origin of the slight departures from perfect isotopic spin symmetry. The masses of particles within a given family are not precisely equal, generally differing by less than 1 percent to several percent. (The masses of the best-known family pair, the neutron and the proton, differ by only .13 percent.) The differences in mass are about what one would expect if isotopic spin symmetry were respected by the strong interactions but violated by the electromagnetic ones. Calculations along these lines, however, never seem to work. For instance, the electromagnetic self-energy of the proton not

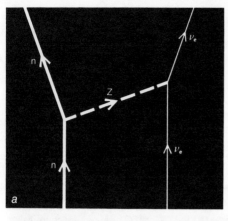 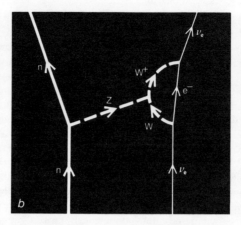 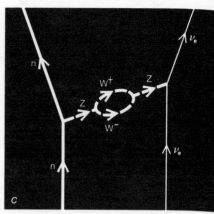

POSSIBLE SOLUTION OF INFINITY PROBLEM in calculating rates of neutron-neutrino scattering may be achieved by postulating the existence of the Z particle, a neutral intermediate vector boson. Such a particle is predicted by unified theory of weak and electro-magnetic interactions proposed by the author. The Z particle should lead to a variety of neutrino-scattering events of the kind diagrammed here. When such processes are added to those involving the charged intermediate vector boson, $W^\pm$, it is found that the

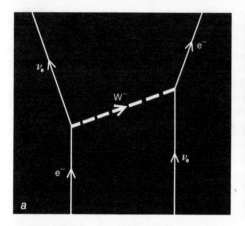 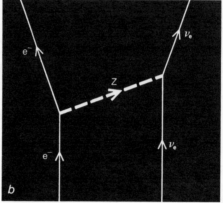 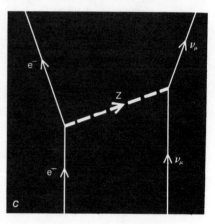

TESTS FOR EXISTENCE OF Z PARTICLE can be made by studying interactions of neutrinos with electrons or with protons and neutrons. The scattering of an electron-type neutrino, ν_e, by an electron can involve the exchange of either a charged intermediate vector boson $W^\pm$ (a) or a neutral intermediate vector boson Z (b). Hence the process can be used to test for the Z particle only if the rate of the process is carefully measured and compared with theory. In contrast the scattering of a muon-type neutrino, ν_μ, by an electron (c) or of any kind of neutrino by a proton or neutron (d) can occur only by exchange of a Z particle and therefore provides

only turns out to be positive, contrary to the observation that the neutron is slightly heavier than the proton, but also has an infinite value. This infinity is of the type discussed above, but it cannot be eliminated by renormalization of the bare mass of the proton, if we insist that the bare masses of the proton and the neutron are equal.

If, as now seems possible, the weak interactions really have an intrinsic strength comparable to that of the electromagnetic interactions, they can provide additional corrections to isotopic spin symmetry that can cancel the infinities due to electromagnetism and leave a finite correction of the right magnitude and sign. Before such calculations can be effectively carried out, however, it is necessary to settle on a detailed model not only of the weak interaction of electrons and neutrinos, as in the 1967 theory, but also of the weak interactions of the strongly interacting particles. This task is still in progress.

Since the weak and electromagnetic interactions seem to be described by a unified gauge-symmetric field theory, it is natural to ask whether the strong interaction can be brought into this picture. There are in fact good reasons for seeking a description of strong interactions in terms of gauge field theories. Possibly the most important is that for a certain class of such theories it is possible to prove that the strong and electromagnetic interactions must necessarily exhibit symmetries between right and left and between matter and antimatter, as is in fact observed to be the case, even though these symmetries are not respected by the weak interactions. As we have seen, the difficulty in testing such field theories is not the lack of experimental data but rather the lack of a method of calculation that can cope with the strength of the strong interactions. Within the past year, however, there has been a theoretical breakthrough that may at last make possible a solution of

this problem. David Politzer, a graduate student at Harvard, and independently David Gross and Frank Wilczek of Princeton University, have discovered that in certain gauge field theories the effective strength of the strong interactions at a given energy decreases as the energy rises. In such "asymptotically free" theories it is possible to carry out approximate calculations with the same methods one uses for the weak and electromagnetic interactions, provided that one works at an energy sufficiently high (no one really knows how high) for the strong interactions to be sufficiently weak. Some of the calculations carried out in this way seem to agree quite well with experiment and others do not.

Although it is too early to tell how this will all work out, the development of asymptotically free gauge theories has already led Gross, Wilczek, Politzer, Georgi, Glashow, Helen Quinn and me to an intriguing series of conjectures. If the effective interaction strength be-

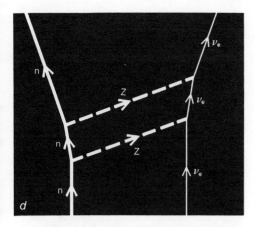

infinities appearing in the sum of the contributions to the total rate, symbolized by all such diagrams, can be absorbed into a renormalization of parameters of new theory.

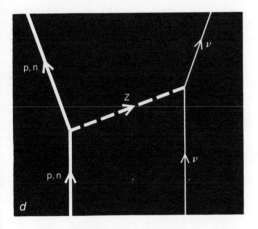

direct evidence for a new kind of weak interaction. These interactions are neutral-current processes such as the one shown in the bubble-chamber picture from Argonne National Laboratory at top right on page 111.

comes small at high energies and short distances, then it must become large at low energies and large distances. Perhaps this explains why the ordinary elementary particles cannot be broken up into quarks: as a quark is pulled away from the rest of the particle the forces may increase without limit. Perhaps the intrinsic interaction strength of the strong interactions is really of the same order of magnitude as that of the weak and electromagnetic interactions and only appears stronger because our present experiments happen to be carried out at relatively low energies and large distances. Perhaps the strong interactions are really caused by the exchange of particles that belong to the same family as the photon and the intermediate vector bosons that are responsible for the electromagnetic and weak interactions. If these speculations are borne out by further theoretical and experimental work, we shall have moved a long way toward a unified view of nature.

Supergravity and the Unification of the Laws of Physics

by Daniel Z. Freedman and Peter van Nieuwenhuizen
February, 1978

In this new theory the gravitational force arises from a symmetry relating particles with vastly different properties. The ultimate result may be a unified theory of all the basic forces in nature

A catalogue of the most basic constituents of the universe would have to include dozens or even hundreds of particles of matter, which interact with one another through the agency of four kinds of force: strong, electromagnetic, weak and gravitational. There is no obvious reason why nature should be so complicated, and perhaps the most ambitious goal of modern physics is to discover in the diversity of particles and forces a simpler underlying order. In particular, a more satisfying understanding of nature could be achieved if the four forces could some-

how be unified. Ideally they would all be shown to have a common origin; they would be viewed as different manifestations of a single more fundamental force.

In the past 50 years remarkable progress has been made in identifying the elementary particles of matter and in understanding the interactions between them. Of course, many problems remain to be solved; two of the most fundamental ones concern gravitation. First, it is not understood how gravitation is related to the other fundamental forces. Second, there is no workable theory of

gravitation that is consistent with the principles of quantum mechanics. Recently a new theory of gravitation called supergravity has led to new ideas on both these problems. It may represent a step toward solving them.

Of the fundamental forces in nature gravitation was the first to be recognized and the first for which a mathematically accurate theory was found, namely the theory published by Newton in his *Principia* of 1687. Newton devised the simple law that the gravitational force acts universally between all pairs of particles with a strength directly proportional to the product of their masses and inversely proportional to the square of the distance between them. He could then calculate both the motions of terrestrial projectiles, which agreed with the observations of Galileo, and the orbits of the planets, which agreed with the empirical laws of planetary motion formulated by Kepler. The development of a law of force that correctly described both terrestrial and astronomical motions was an extraordinary synthesis.

A similar unification was accomplished in the treatment of electromagnetism. To the natural philosophers of the 18th century there was no apparent relation between the static electricity generated by combing one's hair, the magnetic force on a compass needle and the light emitted by a candle or the sun. In the 19th century, however, all these phenomena were shown by James Clerk Maxwell to be related by a set of differential equations, which are now known as Maxwell's equations for the electromagnetic field.

Near the end of the 19th century the opinion prevailed that all the complex manifestations of gravitation and electromagnetism could be described by the laws of Newton and Maxwell. All that remained was to work out the consequences of the equations in detail. This

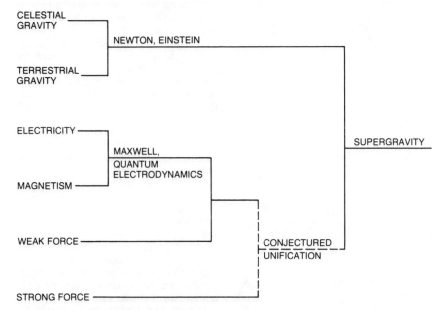

HISTORY OF THEORIES in physics suggests a gradual development toward unification. The first great synthesis was achieved by Newton, who showed that the motion of projectiles on the earth and the orbits of the planets could be explained by the same simple law. In a similar way James Clerk Maxwell devised a single theory encompassing both electricity and magnetism. In the 20th century Newton's theory has been superseded by Einstein's general theory of relativity, and Maxwell's theory has been extended to create a quantum field theory, called quantum electrodynamics. Recently the weak force has been combined with electromagnetism. The strong force is also described by a quantum field theory, and ultimately it may be possible to regard all three of these forces (strong, weak and electromagnetic) as manifestations of a single principle. Supergravity is a theory developed in the past two years that describes gravitation in terms of a quantum field theory. It is still untested but it might unify all forces in nature.

complacent view was shattered by a series of experimental results obtained in the decades immediately before and after the turn of the century. One important conflict between theory and experiment was created by the discovery that the speed of light, unlike the speed of all other waves, does not depend on the motion of the observer. Another difficulty was the interpretation of the discrete lines in the spectra of atoms. It had long been known that atoms emit light only at particular, characteristic frequencies, but this observation could not be reconciled with the discovery that atoms consist of electrons in orbit around a tiny, dense nucleus. Maxwell's theory predicted that a continuous spectrum of light would be emitted as electrons spiral in toward the nucleus.

These conflicts were resolved with the development of two theories that have become foundation stones of modern physics: the special theory of relativity and quantum mechanics. In order to achieve this resolution it was necessary to abandon the concepts of absolute time and determinism in the motion of particles, concepts that were so deeply ingrained it was difficult to recognize that they were actually hidden assumptions. In special relativity time and space were related, and in quantum mechanics particles and waves were shown to be equivalent. It was then possible to understand why the speed of light is the same for all observers and why the spectral lines of atoms have fixed, discrete frequencies.

In Newton's theory of gravitation space and time do not have the close relation they have in special relativity, and so special relativity made necessary a revised theory of gravitation. Such a theory was proposed by Einstein in 1916 and was called the general theory of relativity. It too became a foundation stone of modern physics.

QUANTUM FIELD THEORIES explain the forces acting on a particle in terms of other particles that can be emitted or absorbed. Such events are represented graphically by vertexes. In the basic vertex of quantum electrodynamics lines representing an electron (e^-) or a positron (e^+) intersect a third line representing a photon (γ). The vertex can be interpreted three ways, depending on the directions in which the various particles are moving. An electron and a positron can annihilate each other to yield a photon (*left*); a photon can decay into an electron and a positron (*middle*), or an electron can decay into a photon and another electron (*right*). The creation and annihilation of particles and antiparticles is the characteristic process that distinguishes quantum field theories from "classical" field theories such as Maxwell's or Einstein's.

Early in the 20th century two new fundamental forces of nature were discovered. They are the weak interaction, which is responsible for the beta decay of radioactive elements, and the strong force, which binds together protons and neutrons in atomic nuclei. These forces had not been discovered earlier because they are effective only over the short range of subatomic distances, whereas gravitation and electromagnetism are long-range forces that can be observed macroscopically.

The four forces present a bewildering variety of properties. Differences in range have already been mentioned; there are even greater disparities in effective strength. As might be expected, the strong force is the most powerful of the four (at short range). If the strong interaction between two protons is defined as having a strength of 1, then the electromagnetic force between the same particles has a strength of about 10^{-2} and the weak force has a strength of 10^{-5}. The gravitational force is extraordinarily feeble, with an intrinsic strength of only 10^{-39}. One can visualize how weak that force is by imagining an atom in which the electrons are bound to the nucleus by gravitation instead of electromagnetism; a single hydrogen atom would then be far larger than the estimated size of the universe.

In spite of the disparate properties of the four forces it is natural to search for a deeper theory in which they would have a common origin. Einstein devoted much of the last part of his life to the search for a unified field theory of gravitation and electromagnetism. Other physicists joined the effort, but the results of their work were never convinc-

INTERACTIONS BETWEEN PARTICLES can be visualized by drawing diagrams in which two vertexes are connected. The force between the two particles is transmitted by the exchange of a third particle, which is said to be virtual because it cannot be detected. Each of the four forces in nature has its own virtual particles, or quanta. The quantum of the electromagnetic force is the photon. Strong interactions between quarks (the supposed constituents of protons and neutrons) are mediated by particles called gluons. The weak force is transmitted by particles (W^+, W^- and Z^0) that are thought to be exactly like the photon except that they have a large mass. The quantum of gravitation is the graviton, a massless particle. Of these exchanged quanta only the photon has been observed, but there is confidence that the others also exist.

ing, perhaps because quantum-mechanical ideas were not incorporated in their theories.

During the past decade the weak and the electromagnetic forces have been unified in the work of Steven Weinberg of Harvard University and later in that of Abdus Salam of the International Centre for Theoretical Physics in Trieste and John Ward of New Zealand. It is thought that a similar approach can also encompass the nuclear force. Ironically, however, this synthesis does not include the force that has been known the longest: gravitation.

The theory of supergravity suggests a new approach to unification. Supergravity is an extension of general relativity, and it makes the same predictions for the classical tests of Einstein's theory, such as the precession of planetary orbits, the bending of starlight passing near the sun, the red shift of stellar spectral lines and the delay of radar signals passing through the solar gravitational field. At the microscopic, or quantum, level, however, supergravity is different from general relativity. When the probability of certain quantum-mechanical effects of gravitation is calculated in general relativity, the probability turns out to be infinite, a result that makes no sense. In supergravity finite answers have been obtained in all calculations that have been done so far.

In constructing new physical theories it is helpful to be guided by principles of symmetry, which allow one law to describe objects or concepts that had seemed unrelated. As will be explained below, a symmetry of a physical theory can hold in either global or local form. Theories with local symmetry, which are also called gauge theories, have proved to be much more powerful. The

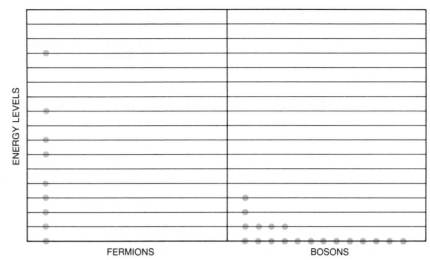

SPIN ANGULAR MOMENTUM affects not only the kinetics of particles but also the statistical behavior of systems made up of two or more identical particles. Those particles whose spin is half an integer (such as the electron and the proton) are called fermions. They obey the exclusion principle formulated by Wolfgang Pauli: No two fermions can occupy the same quantum-mechanical state. Particles with integer spin (such as the photon and the graviton) are called bosons, and they can be brought together in unlimited numbers at one point or in one quantum-mechanical energy state. Hence in any system with a spectrum of energy states fermions distribute themselves one to a state, whereas bosons tend to condense in the lowest state available.

general theory of relativity and Maxwell's theory of electromagnetism are both based on local symmetries. The recent unified field theories of the weak and the electromagnetic interactions are also gauge theories. It is therefore suspected that any theory unifying the four forces should also have local symmetry.

Supergravity is based on a new symmetry so remarkable even at the global level that it has been given the name supersymmetry. Supersymmetry relates the two broad classes of elementary particles, namely the fermions (such as the electron, the proton and the neutron) and

the bosons (such as the photon). Fermions and bosons have vastly different properties, and finding a fundamental relation between them was quite unexpected. In supergravity, supersymmetry is extended from the global level to the local level. Remarkably, this extension leads automatically to theories that incorporate the gravitational force and suggest the possibility of unifying gravitation with the other forces.

Supergravity has not been tested by experiment. It is still a speculative theory, but the progress that has been made so far is encouraging. By bringing to-

FORCE	STRENGTH	RANGE	PARTICLES ACTED ON	PARTICLES EXCHANGED	MASS OF EXCHANGED PARTICLES	SPIN OF EXCHANGED PARTICLES	NATURE OF FORCE BETWEEN IDENTICAL PARTICLES
STRONG	1	SHORT	QUARKS	GLUONS	?	1	REPULSIVE
ELECTRO-MAGNETIC	10^{-2}	LONG	ELECTRICALLY CHARGED PARTICLES	PHOTONS	0	1	REPULSIVE
WEAK	10^{-5}	SHORT	ELECTRONS, NEUTRINOS AND QUARKS	INTERMEDIATE VECTOR BOSONS (W^+, W^-, Z^0)	50–100 GeV	1	REPULSIVE
GRAVITATIONAL	10^{-39}	LONG	ALL PARTICLES	GRAVITONS	0	2	ATTRACTIVE

FOUR FORCES have a character determined largely by the properties of the associated quanta. The mass of the exchanged particle determines the range of the force; only forces transmitted by massless particles can have a long range. Spin angular momentum also has an important influence. Forces transmitted by particles whose spin is an even integer are invariably attractive; quanta with odd-integer spin give rise to repulsive forces between like particles. Although gravitation is exceedingly feeble, it is the only one of the four forces that is both long-range and attractive between all pairs of particles. Gravitation therefore determines the large-scale structure of the universe.

gether several of the most fundamental concepts in modern physics, supergravity has already achieved more than any earlier quantum theory of gravity.

The present understanding of the fundamental laws of nature arose from three principles: special relativity, general relativity and quantum mechanics. Each of them resolved an experimental or theoretical conflict and each went on to predict new phenomena that were subsequently verified by experiment. Today there can be little doubt about their validity.

Einstein proposed the special theory of relativity in order to reconcile the concept of motion in Newtonian mechanics with the experimental finding that the speed of light is constant for all observers. If light waves behaved in the same way that waves on the surface of water behave, an observer in motion with respect to the waves would mea-

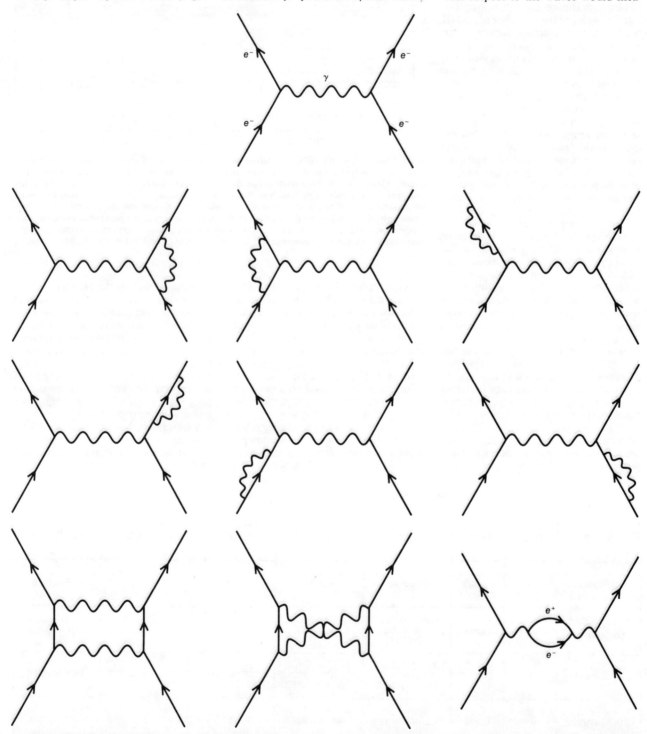

QUANTUM CORRECTIONS to the classical laws of force are represented by diagrams that have closed loops. In the interaction between two electrons the exchange of a single virtual photon (*top*) corresponds to the force predicted by Maxwell's theory of electromagnetism. In quantum electrodynamics more complicated interactions must be considered. For example, a photon can be emitted and then reabsorbed by the same electron, two photons can be exchanged, or the virtual photon can give rise to an electron and a positron, which then annihilate each other to yield another photon. A long-standing problem in quantum field theories is that such loop diagrams predict an infinite probability of interaction. In quantum electrodynamics a procedure called renormalization eliminates the infinities. The difficulty of formulating a quantum theory of gravity is that the renormalization procedure is not effective and some infinities persist.

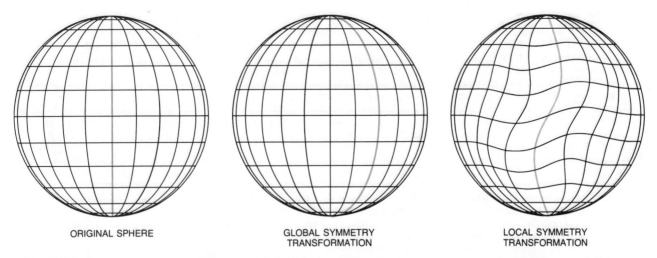

ORIGINAL SPHERE   GLOBAL SYMMETRY   LOCAL SYMMETRY
                                TRANSFORMATION   TRANSFORMATION

SYMMETRIES can be present in laws of nature as well as in patterns and in objects such as crystals. Just as a crystal retains its form after a specified rotation or translation, a symmetrical law of nature remains invariant after some specified transformation. The symmetries in physics are of two types, called global and local; the distinction between them can be illustrated by considering an ideal spherical balloon (*left*), marked with a system of coordinates so that the positions of all points on the surface can be identified. A global symmetry is exhibited if the sphere is rotated about some axis (*middle*). The rotation is a symmetry operation because the form of the sphere remains unchanged; it is a global symmetry because the positions of all the points on the surface are changed by the same angular displacement. Local symmetry requires that the balloon keep its shape even if the points on the surface are moved independently (*right*). It is notable that a local symmetry operation stretches the balloon and therefore introduces forces between points. Each of the four basic forces in nature is thought to arise from a similar requirement that a law of nature be invariant under a local symmetry transformation.

sure a wave velocity different from that measured by an observer at rest. Maxwell's equations seemed to predict that the speed of light did not depend on the motion of an observer, and since the prediction was in conflict with common sense it was supposed Maxwell's equations were valid only for observers at rest. In 1888 the experiment of A. A. Michelson and E. W. Morley showed that this supposition was false. It was Einstein's genius to realize that the elegant form of Maxwell's equations was more important than the Newtonian, or commonsense, view of motion. He showed that Maxwell's equations become valid for all observers if the scales of both length and time employed by any two observers depend intrinsically on their relative velocity. He then showed that the laws of Newtonian mechanics could be modified to incorporate these new concepts of space and time. One important prediction is that the speed of light is the maximum velocity allowed for any particle or signal. Another well-known prediction is the relation between the energy and mass of a particle, $E = mc^2$, a formula that determines the energy released in nuclear reactions.

The general theory of relativity was proposed by Einstein in 1916, after nine years of grappling with the problem of formulating a theory of gravitation in agreement with the space-time symmetry of special relativity and with the experimental observation, known since Galileo, that all bodies, regardless of their mass, follow the same trajectory in a gravitational field. In its conceptual content and in the breadth of creative insight required, general relativity is a stunning achievement of the human intellect. Roughly speaking, Einstein first reasoned that if the trajectories of falling bodies are independent of their mass, gravitation must be related to the intrinsic structure of space and time. He then derived equations that express this idea in mathematical form, drawing on the methods of non-Euclidean geometry formulated a century earlier by Carl Friedrich Gauss and Bernhard Riemann. The resulting theory predicted that starlight passing near the limb of the sun would be bent toward the sun twice as much as Newton would have predicted. Einstein suggested that this effect be measured during a total eclipse of the sun. The results of an expedition led by A. S. Eddington were announced on the first anniversary of the armistice of World War I. It was a dramatic and sublime moment in the history of science. Einstein's deep theoretical reasoning had found confirmation in the ultimate laboratory of nature.

Years after the formulation of general relativity Einstein wrote: "In the light of the knowledge obtained, the happy achievement seems almost a matter of course and any intelligent student can grasp it without too much trouble. But the years of anxious searching in the dark with their intense longing, their alternations of confidence and exhaustion and the final emergence into light—only those who have experienced it can understand that." His words apply equally well to other pathbreaking discoveries in physics.

Quantum mechanics, which was developed in 1926 by Werner Heisenberg, Erwin Schrödinger and others, gave a convincing explanation of the discrete lines in atomic spectra. Quantum mechanics describes the electrons in atoms not as point particles but as superpositions of waves, which can be interpreted as a probability distribution around the nucleus. The energy of each distribution has a definite value, and radiation at discrete frequencies is emitted when an electron jumps from one such quantum state to another. The strict determinism of classical mechanics is abandoned in the quantum theory and is replaced by a probabilistic interpretation of measurements at the microscopic level. For example, an electron in an atom can be anywhere around the nucleus; the probability of finding it at any given point is related to the amplitude of the wave distribution at that point.

If modern theoretical physics was to be built on the foundations of special relativity and quantum mechanics, it was essential to bring the two theories together. The first major advance was made by P. A. M. Dirac, who formulated a relativistic quantum wave equation for the electron in 1928. A surprising prediction emerged from Dirac's work. His equation was consistent only if there existed a new particle with the same mass as the electron but opposite electric charge. This antiparticle of the electron, called the positron, was discovered in 1932. It is now recognized that there must be an antiparticle for every particle in nature.

A full unification of special relativity and quantum mechanics came with the development of quantum field theory, which began with work by Dirac, Heisenberg and Wolfgang Pauli in the late 1920's. Quantum field theory is a general formalism that in principle can be applied to each of the four forces. In practice, however, difficulties arise from the infinities that appear in the calculation of certain quantum contributions to probabilities. These difficulties were first overcome in quantum electrodynamics, which is the quantum field theory that describes the electromagnetic interactions of electrons, positrons and photons. Success came in the late 1940's, two decades after the formulation of the theory, when Richard P. Feynman, Julian S. Schwinger, and Sinitiro Tomonaga developed simplified methods of calculation in closer harmony with the underlying symmetries of the theory. It was found that through a method called renormalization the infinities could be removed in an unambiguous way. The finite predictions obtained could then be compared with experiment. The predictions of quantum electrodynamics are confirmed with extraordinary accuracy. For example, the theory predicts that the electron acts as a tiny magnet. The measured value of the electron's magnetic strength, which is $1.0011596524 \pm .0000000002$, agrees with the theoretical prediction to within a few parts in 10 billion.

The renormalization procedure can be applied only in a special class of quantum field theories where the infinities can be compensated for by correction of the basic parameters of the theory, such as the mass and the charge of the electron. The observed electron mass is the sum of the "bare mass" and the "self-energy" resulting from the interaction of the electron with its own electromagnetic field. Only the sum of the two terms is observable. The self-energy can be calculated and turns out to be infinite. Nothing is known about the bare mass, and so it can be assigned a negatively infinite value, with the result that the two infinities cancel and yield the observed finite mass of the electron.

Renormalization works in quantum electrodynamics and in the currently accepted field theories constructed to describe the strong interactions. For many years it seemed there was no convincing renormalizable theory of the weak interactions. That situation was changed in the early 1970's by a theoretical discovery made by Gerard 't Hooft and Martin J. G. Veltman of the University of Utrecht, which was developed further by them and by the late Benjamin W. Lee of the Fermi National Accelerator Laboratory and by J. Zinn-Justin of the Saclay Nuclear Research Center near Paris. They showed that the unified field theories of the weak and the electromagnetic interactions can be renormalized. Quantum theories of the gravitational force still have serious difficulties with infinities; it is here that supergravity offers new hope.

In quantum field theory particles are identified with waves or quanta of the underlying field. An important prediction of quantum field theory is that particles spin about an axis. The rate of spin can take only values that are integer or half-integer multiples of Planck's constant. The short-lived pi meson has a spin of 0, the electron, the proton and the neutron have a spin of 1/2 and the photon has a spin of 1.

One general prediction of quantum field theory that is strikingly confirmed in nature is the connection between the spin of a particle and its statistics, that is, the behavior of a system made up of two or more identical particles. Particles with half-integer spin (such as 1/2, 3/2, 5/2) are fermions, and they obey the exclusion principle formulated by Pauli, which states that no two identical fermions can occupy the same point in space or, more generally, the same quantum state. It is this property of fermions that explains the elaborate structure of electron shells in atoms and hence the diverse chemical properties of the elements. If it were not for the exclusion principle, all the electrons in an atom would condense in the lowest energy level. In bulk matter at high density the same principle leads to an effective repulsive force between identical fermions. This force explains the stability of white-dwarf stars and neutron stars, which without it would collapse into black holes under the attractive gravitational force.

For particles with integer spin, the bosons, the statistics are entirely different. There is an increased probability for two or more bosons to occupy the same point in space or the same quantum state. Such superpositions of many identical bosons can lead to macroscopically observable effects. For example, laser

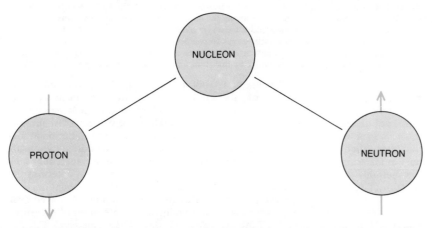

ISOTOPIC SYMMETRY establishes a relation between particles with the same spin angular momentum, such as the proton and the neutron. Both these particles can be regarded as states of a primitive or undifferentiated particle called the nucleon, which can be imagined to have an arrow associated with it in some fictitious space. If the arrow points up, the nucleon is a proton; if it points down, the nucleon is a neutron. For real particles the arrow is either up or down, since no real particle is half proton and half neutron, but the laws of physics that describe interactions between protons and neutrons remain invariant under arbitrary rotations of the arrow.

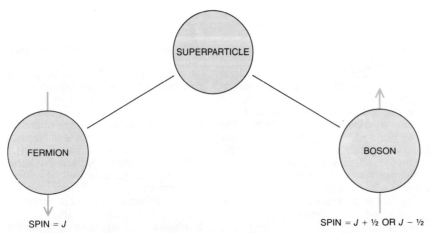

SUPERSYMMETRY relates particles with different spins, namely those with the adjacent spins of J and either $J + 1/2$ or $J - 1/2$. Hence any fermion and boson with adjacent spins can be regarded as alternative manifestations of a single "superparticle" with an arrow in an auxiliary space. Such a fundamental symmetry between fermions and bosons was long thought to be impossible. In quantum field theories with local supersymmetry the gravitational force appears naturally. The spin-2 graviton has a fermion partner with a spin of 3/2. The name gravitino has been suggested for this particle. Neither the graviton nor the gravitino has been observed.

light is a superposition of many photons with the same energy and direction. Given the disparate properties of bosons and fermions, it is all the more remarkable that a symmetry has been found that can relate them.

In quantum field theory all forces between particles are described by the exchange of "virtual" particles. The repulsive force between two electrons comes about when one electron emits a photon, which is then absorbed by the other electron. The intermediate photon is said to be virtual because it cannot be observed directly in this process; it exists for too short a time. It can be proved in quantum field theory that long-range forces arise from the exchange of massless particles and short-range forces arise from the exchange of particles with mass. The short-range nuclear force between two protons arises from the exchange of pi mesons, whose mass is 300 times that of an electron. At a more fundamental level the proton, the neutron and the pi meson are thought to be made up of quarks. In this view the strong force arises from the exchange by the quarks of virtual particles called gluons. The weak force is thought to arise from the exchange of intermediate vector bosons, which are very heavy particles of spin 1 predicted by the unified field theories of weak and electromagnetic interactions.

Although a satisfactory theory of quantum gravity is yet to be established, one can anticipate that the Newtonian force will arise from the exchange of a virtual particle. This particle is called the graviton. It must be massless because gravity is a long-range force. Its spin must be an even integer, such as 0, 2 or 4, because it can be shown that the exchange of bosons whose spin is an odd integer gives rise to forces that are repulsive between like particles. The spin

cannot be 0 because in that case light would not bend toward the sun. The next-simplest case, spin 2, satisfies all experimental tests.

The arguments of the preceding paragraph can be reversed, that is, the general theory of relativity can be derived from the assumption that the gravitational force results from the exchange of massless spin-2 particles and that these particles are described by a quantum field theory. All particles are pulled by the virtual gravitons in such a way that they follow the same curved trajectories predicted by Einstein's equations. This new derivation of general relativity does not detract from Einstein's discovery. It simply illuminates it from the new viewpoint of forces in quantum field theory.

The basic difference between classical field theories and quantum field theory is that only in the latter can particles be created and destroyed. The annihilation of an electron and a positron to yield two photons is possible only in quantum field theory. Since the process is frequently observed in the laboratory, there is no doubt that this aspect of quantum field theory is correct. Indeed, the prediction of the probability for this event by quantum electrodynamics is in complete agreement with experiment.

The basic element of a quantum field theory is the "vertex," in which one particle disintegrates to yield two or more other particles. The vertex for quantum electrodynamics is the intersection of two lines that represent electrons or positrons and one line that represents a photon. The vertex can describe the creation of an electron-positron pair by a decaying photon, the annihilation of an electron-positron pair creating a photon or the decay of an electron into a second electron and a photon.

The probability for any process in a quantum field theory is obtained by forming diagrams in which vertexes are

connected in all possible ways. For the scattering of one electron by another the simplest diagram shows the exchange of a single photon, which represents the effects of the classical electromagnetic force. Other diagrams have closed loops of virtual particles. The loop diagrams represent the quantum corrections to the classical law of force. For each diagram there is a well-defined but complicated mathematical expression. In successful quantum field theories it is possible to evaluate these expressions and to remove the infinities they often exhibit by renormalization. The probability for the process is then obtained by adding the amplitudes for all the diagrams and squaring the sum. There are an infinite number of diagrams and so it is impossible to obtain an exact result. In quantum electrodynamics the contribution of the diagrams with one loop is smaller than the contribution of the diagrams with no loops by a factor of less than 1 percent. Diagrams with two loops are smaller by an additional factor of 1 percent, and so on. Hence an accurate approximation of the total amplitude can be obtained by considering only a few diagrams.

Throughout history the decorative patterns of many cultures bear witness to the strong appeal symmetry has always had for mankind. A symmetry can be understood as a motion that leaves the form of a pattern or an object invariant. For example, a cube rotated 90 degrees appears to be unchanged, and a sphere is invariant after any rotation around its center. Physical theories can have symmetries of a similar kind, but what is invariant after a transformation is not a pattern or an object but the mathematical laws of the theory itself. Physicists now recognize that symmetries play a vital role in our understanding of nature.

As an example of a fundamental sym-

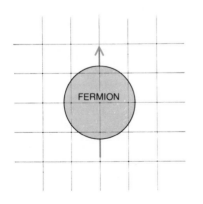 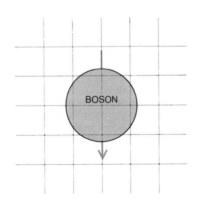 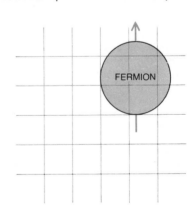

FERMION ——— SUPERSYMMETRY ———→ BOSON ——— SUPERSYMMETRY ———→ FERMION
SPIN = J   TRANSFORMATION   SPIN = J + ½ OR J − ½   TRANSFORMATION   SPIN = J

SUPERSYMMETRY TRANSFORMATIONS result in a change in the position of a particle. Supersymmetry seems to be an internal symmetry that concerns only the properties of a particle and not its position. Remarkably, however, a repeated supersymmetry transformation, such as one from fermion to boson and back to fermion, moves a particle from one point in space to another. In local supersymmetry the displacement can be different at each point in space. The displacement of a particle through a supersymmetry transformation suggests a relation between supersymmetry and the structure of space-time; it is this relation that gives rise to the gravitational force.

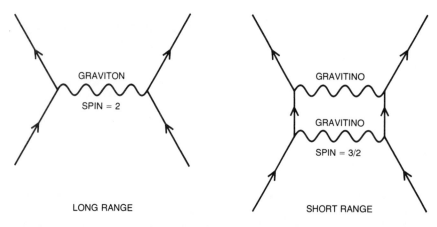

GRAVITON
SPIN = 2

LONG RANGE

GRAVITINO

GRAVITINO

SPIN = 3/2

SHORT RANGE

SUPERGRAVITY provides corrections to the general theory of relativity at the quantum level. In general relativity the gravitational force arises entirely from the exchange of gravitons; in supergravity there is an additional contribution from the exchange of spin-3/2 gravitinos. Because the gravitinos are fermions they are exchanged only in pairs, a process with negligible probability except at short range. The predictions of general relativity for long-range interactions are unchanged in supergravity; new effects are predicted only at microscopic scale.

metry, consider two astronauts studying electromagnetic phenomena from two spacecraft drifting freely in some interstellar region, where they would have constant relative velocity. Each astronaut defines a coordinate system with the origin at his own position and with different orientation. The two astronauts therefore identify external events by different coordinates, and records of their observations would look quite different. If they reduce their findings to physical laws, however, they both find that Maxwell's equations are valid.

The principle demonstrated by this thought experiment is called Poincaré invariance, after the French mathematician Henri Poincaré. It is the space-time symmetry underlying the special theory of relativity. Comparison of the observations made by the two astronauts would show that their coordinate systems are related in the manner required by special relativity. Poincaré invariance states explicitly the experimentally verified assumption that the laws of nature are the same across town as they are in your backyard. More precisely, it states that all laws of physics take the same form in any two coordinate systems, even if they are shifted and rotated and moving with respect to one another. as long as they have a constant relative velocity.

There is a vital distinction between symmetries such as this one. which are said to be global, and local symmetries. Global symmetry may sound like the grander concept, but local symmetries impose more stringent requirements on theories and reveal deeper unities in nature. Indeed, the transition from a global symmetry to a local one describes the origin of the gravitational and the electromagnetic forces, and there is reason to suspect that the other forces also emerge from local symmetries.

A global symmetry is one in which a

transformation is applied uniformly to all points in space; in a local symmetry each point is transformed independently. The distinction can be illustrated by means of a spherical balloon with various points marked on its surface. If an axis through the center of the sphere is chosen, the balloon can be rotated through some angle. Because the balloon keeps its spherical form the rotation is a symmetry operation of the balloon; because all the points on the surface are transformed in the same way (by rotation through the same angle) it is a global symmetry operation.

With an ideal balloon it would also be possible to move each point independently, pushing or pulling the points to new positions on the surface but keeping their distance to the center fixed. Again the sphere retains its form, so that this procedure is also a symmetry operation. Each point is transformed independently of its neighbors and so it is a local symmetry. There is now a significant change: when the points are moved independently, the membrane of the balloon is stretched and elastic forces develop between the displaced points. These forces are not artifacts of an imperfect model; on the contrary, forces appear in the same way when a physical theory has a local symmetry.

The Poincaré invariance discussed above is a global symmetry because a transformation between the two sets of coordinates employed to describe a given point in space-time is the same for all points. The much stronger constraint of local Poincaré invariance is established by requiring the laws of physics to retain the same form when the coordinates of each point are transformed independently. This change is equivalent to allowing the two observers to have accelerated motion with respect to each other. At first it would seem that observers under these circumstances would not

derive the same laws of physics because an accelerating observer experiences "fictitious" forces such as the centrifugal force of rotational motion. Einstein recognized that fictitious forces induced by acceleration are closely related to gravitational forces associated with masses. He showed that the laws of physics can be kept invariant if the gravitational field is introduced into the equations. The result is the general theory of relativity.

The above example illustrates a powerful, general feature of the relation between global and local symmetries. If a set of physical laws is invariant under some global symmetry, the stronger requirement of invariance under local symmetry can be met only by introducing new fields, which give rise to new forces. The fields are called gauge fields, and they are associated with new particles whose exchange gives rise to the corresponding forces. Thus gravitation is the gauge field of local Poincaré invariance, and the gravitational force results from the requirement that Poincaré symmetry be local.

The existence of electromagnetic forces can also be derived from the requirement of local symmetry. In a quantum field theory charged particles are described by fields that have two numbers at every point in space-time: an amplitude and a phase. The amplitude measures the probability of finding a particle at a point and the phase describes the wave properties of the particle. Observable quantities such as the total energy of a set of charged particles do not change when the phase of the field is shifted by an amount that is the same at all points. Thus the field has a global symmetry under a change of phase. Local symmetry would require that observable quantities remain invariant when the phase is allowed to vary independently at each point. To accommodate the local symmetry it is necessary to introduce the electromagnetic field as a gauge field; the quanta of this field are photons, which give rise to the electromagnetic force. With only a global symmetry there would be no electromagnetic forces between charged particles, no photons and no light.

Local electromagnetic symmetry is called an internal symmetry; unlike Poincaré invariance it does not involve a change in space-time coordinates. Another internal symmetry, called isotopic symmetry, is fundamental to nuclear physics; it establishes a relation between the proton and the neutron.

The proton and the neutron both have a spin of 1/2, and they experience nuclear forces of about the same strength. Isotopic symmetry allows them to be regarded as two alternative states of a single undifferentiated particle, the nucleon. The nucleon can be imagined as a particle with an arrow in some imaginary space. If the arrow points up, say, the particle is a proton; if it points down, it is a neutron. Physical particles always have the arrow pointing either up or down (there are no real particles that are half proton and half neutron), but the equations that describe the nuclear force are invariant under arbitrary rotations of the arrow. In nuclear physics isotopic symmetry is a global symmetry. The arrow must rotate through the same angle at all points if the nuclear forces are to be invariant.

The problem of extending isotopic symmetry from a global symmetry to a local one was solved in 1954 by C. N. Yang and Robert Mills. They found that the transition to a local symmetry required the introduction of three gauge fields, each field associated with a massless particle having a spin of 1. For more than a decade the Yang-Mills field theory was regarded as an elegant mathematical curiosity without physical applications. The three massless particles and the long-range forces they would generate simply do not exist.

It is now known that the gauge particles of the Yang-Mills field theory can exist, because they are not necessarily massless. They can acquire a mass—a very large mass—through a mechanism called spontaneous symmetry-breaking. It was shown by Jeffrey Goldstone of the University of Cambridge that the physical manifestations of a symmetrical physical theory can sometimes be quite asymmetrical. An analogy to this process is the roulette wheel: the equations of motion for the ball and wheel are symmetrical around the axis of rotation, but the ball invariably comes to rest in an asymmetrical position.

Peter Higgs of the University of Edinburgh subsequently showed that in theories of gauge fields the effect of spontaneous symmetry-breaking is to give a mass to some of the gauge particles whereas others would be massless. Unified theories of the weak and the electromagnetic forces incorporating local isotopic symmetry and spontaneous symmetry-breaking were then constructed. In these theories the carriers of the short-range weak force are the intermediate vector bosons. They are gauge particles of spin 1, which acquire mass through the Higgs mechanism. The photon is a gauge particle that remains massless. Hence the simple requirement of local isotopic invariance led to the unification of two of the basic forces in nature. As will be shown below, the additional requirement of local supersymmetry can further unify these forces with gravitation.

Internal symmetries such as isotopic symmetry relate particles that have the same spin. One of the dreams of theoretical physics has been to find symmetry schemes that would unite particles having different spins. This dream was realized by the invention of supersymmetry. Supersymmetry relates particles that have adjacent spins, such as 1 and 1/2, and thus of necessity it embraces both fermions and bosons. What is equally remarkable, it relates internal symmetries to Poincaré invariance. It is this connection that allows the construction of the new gravitational theory of supergravity.

Supersymmetry was formulated independently by physicists in the U.S.S.R., western Europe and the U.S. It was discussed in 1971 by Y. A. Golfand and E. P. Likhtman of the Lebedev Physical Institute in Moscow. Their work went unnoticed, however, and the subject was rediscovered in 1973 by D. V. Volkov and V. P. Akulov of the Physical-Technical Institute in Kharkov. A symmetry between bosons and fermions was also found in 1971 by Pierre M. Ramond of

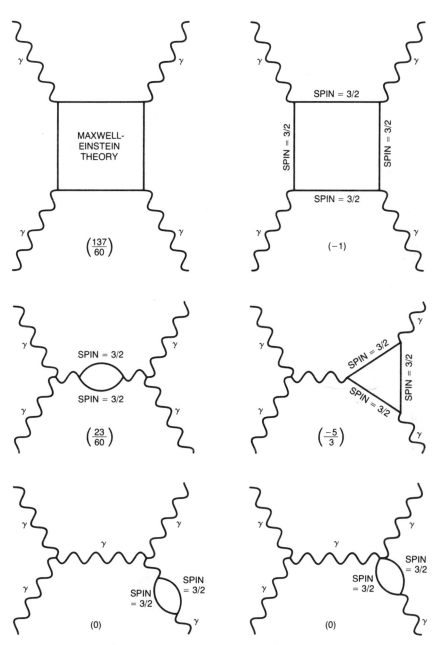

FINITE PROBABILITY for loop diagrams in a quantum theory of gravity is obtained by including gravitinos in the interaction. The diagrams shown here are those for the interaction between two photons. The first diagram, labeled Maxwell-Einstein Theory, consolidates all the one-loop diagrams that involve only gravitons and photons; the contribution from these diagrams is equal to an infinite quantity multiplied by the coefficient 137/60. Five one-loop diagrams involving gravitinos can be constructed; each of them is proportional to the same infinite term multiplied by the coefficients shown in brackets. Only the sum of the diagrams is observable, and adding the coefficients shows that the sum is zero. Hence the infinite contributions of the gravitinos cancel those of the gravitons and the diagrams have a finite probability.

the California Institute of Technology and by André Neveu of the École Normale Supérieure and John H. Schwarz of Cal Tech. Their work grew out of an approach to particle physics called dual models. In 1973 Julius Wess of the University of Karlsruhe and Bruno Zumino of the European Organization for Nuclear Research (CERN) generalized that work to quantum field theory and gave a systematic procedure for constructing global supersymmetry theories.

In the mid-1960's theorems had been proved that seemed to demonstrate the impossibility of the unification of Poincaré invariance and internal symmetry. It is now apparent that these "no go" theorems are incorrect, but the error was not in the proof. Rather, one of the assumptions implicit in the theorems turned out to be too restrictive and could be relaxed. The assumption seems plausible and unexceptional, namely that the numbers employed in describing the symmetry should obey the commutative law for multiplication. It is now known that there is no need for this limitation. Supersymmetry theory incorporates as an essential element numbers that do not have the commutative property.

The commutative law for multiplication states that the product of two numbers does not depend on the order in which they are multiplied. Thus if r_1 and r_2 represent any two real numbers, then $(r_1 \times r_2) - (r_2 \times r_1) = 0$. For example, $(6 \times 2) - (2 \times 6) = 0$. For anticommuting numbers the sign in this equation is changed. If ϵ_1 and ϵ_2 are assumed to be two anticommuting numbers, then $(\epsilon_1 \times \epsilon_2) + (\epsilon_2 \times \epsilon_1) = 0$. Of course, one cannot substitute numerical values in this equation, but one can nonetheless imagine that new numbers with this property exist. Such number systems were invented more than a century ago by Hermann Günther Grassmann, who taught mathematics in the German town of Stettin.

Supersymmetry can be described in much the same way that isotopic symmetry is. A hypothetical "superparticle" is endowed with an extra arrow in an imaginary auxiliary space. If the arrow points up, the particle is a fermion; if it points down, it is a boson. The spins of the fermion and the boson are always adjacent. For example, a boson with a spin of 1 can join a fermion whose spin is either 1/2 or 3/2. For physical particles the arrow must always point either up or down, since no real particles are half boson and half fermion, but the equations that describe the forces between elementary particles are invariant under arbitrary rotations of the arrow, just as they are in isotopic symmetry. The supersymmetry is global if the angle of rotation is the same at all points of space-time and local if the rotation is different at each point.

Let us explain how a supersymmetry transformation works. Denoting the boson and fermion fields b and f, a transformation connects them with new fields b' and f', where $f' = f + b\epsilon$ and $b' = b + f\epsilon$. The factor ϵ is a measure of the angle of rotation of the superparticle arrow. Although b is an ordinary number, f and ϵ are anticommuting numbers.

The necessity of anticommuting numbers in supersymmetry theories is related to the Pauli exclusion principle, the rule that forbids two fermions from occupying the same point. The probability of finding one fermion at a point is given by the value of the field f and the probability of finding two fermions is given by $f \times f$. The Pauli exclusion principle requires that $f \times f = 0$, and anticommuting numbers have precisely this property. The probability of finding two bosons at a point, which is $b \times b$, need not vanish, so that b can be an ordinary number. Compliance with the Pauli principle must also be incorporated in the supersymmetry rotation law. Two bosons can occupy the same point, but if they are both transformed into fermions by a supersymmetry rotation, the probability of finding them at the same point must vanish. The probability is given by $(f + b\epsilon) \times (f + b\epsilon)$, and it must be equal to zero. This requirement implies that $(f \times \epsilon) + (\epsilon \times f)$ is zero and that $\epsilon \times \epsilon$ equals zero.

The most surprising property of supersymmetry is that repeated application of the fermion-boson transformation moves a particle from one point to another in space-time. Thus a Poincaré transformation (of position) is obtained by repeating a supersymmetry transformation. Since local Poincaré invariance is the symmetry that gives rise to general relativity, a connection between supersymmetry and gravitation can also be expected.

It was argued above that the transition from a global to a local symmetry always introduces new gauge fields, which in turn give rise to new forces. An obvious question, then, is whether global supersymmetry can be promoted to a local invariance and, if so, what the nature of the new gauge fields will be. Local supersymmetry is indeed possible if two new fields are introduced; they are the field of the spin-2 graviton and a new spin-3/2 field.

The construction of a gauge theory of supersymmetry begins with the observation that a repeated supersymmetry transformation yields a physical translation of a particle. This relation is embodied in an equation that in effect states that the product of two supersymmetry rotations is a shift in space-time. To obtain a local supersymmetry theory a gauge field must be introduced for each of the symmetries present in the equation. The shift in space-time is a

Poincaré transformation for which the spin-2 graviton is the appropriate gauge particle. Hence gravitation appears naturally in the theory, and it is for this reason that local supersymmetry is usually called supergravity. The gauge field for supersymmetry transformations is less readily determined, but since supersymmetry relates only particles of adjacent spin, it must be a fermion with a spin of 2 + 1/2 or 2 − 1/2, that is, 5/2 or 3/2. Simplicity argues that the spin should be 3/2, and more technical considerations support the same choice.

In 1976 the most elementary example of a supergravity theory was constructed by us in collaboration with Sergio Ferrara of the Frascati Laboratories near Rome. Shortly afterward Stanley Deser of Brandeis University and Zumino showed how to formulate the theory in a simpler way. In our derivation we started with the assumption that the theory should include a spin-2 graviton and a spin-3/2 particle and that it should possess local Poincaré invariance and local supersymmetry. After a great deal of mathematical work we found a theory that meets these requirements.

Work with fields that describe particles of high spin is notoriously difficult. A theory for noninteracting spin-3/2 fields was published in 1941 by Schwinger and William R. Rarita of the Lawrence Berkeley Laboratory of the University of California, but all attempts to introduce forces between the spin-3/2 particles and other particles led to inconsistencies. For example, in some of the proposed theories signals would travel faster than light, with the result that the laws of causality would be violated. We now know that these attempts failed because spin-3/2 particles can be coupled to other particles only when the forces are supersymmetric.

The existence of a fundamental particle with a spin of 3/2 is inescapable in supergravity. The particle is a companion to the graviton, and the name gravitino has been suggested for it. It is not yet clear what properties the gravitino should be expected to have. In the simplest supergravity theories it is massless and is coupled to other particles only by the feeble force of microscopic gravitation. Such particles have not been observed, but they would be exceedingly difficult to detect. Even gravitons have not been observed experimentally. In more complex supergravity theories, however, the gravitino can acquire a mass through spontaneous symmetry-breaking, and stronger forces can arise between the gravitino and other particles. In these theories the gravitino would be much easier to detect. The discovery of such spin-3/2 particles would be strong experimental evidence for supergravity.

Part of the difficulty of uniting gravitation with the other three forces is not that the respective theories are in conflict but that they simply have too little in common. In general relativity forces are derived from the geometric properties of space and time, and in quantum field theory they are derived from the exchange of quanta. Supergravity describes general relativity in the language of quantum field theory, but there is no apparent reason supergravity could not also be formulated in geometric terms. Such a geometric derivation seems to be possible in an extended space-time in which every point has not only the four usual space and time coordinates but also an additional set of coordinates identified by anticommuting numbers. A "superspace" of this kind was introduced by Akulov and Volkov and was investigated further by Salam and John Strathdee of the International Centre for Theoretical Physics. This approach led to the construction of other supergravity theories, which actually preceded the one discussed here. One such theory was developed by Richard L. Arnowitt and Pran Nath of Northeastern University, who followed the same steps as Einstein but in superspace rather than ordinary space. The superspace theories are elegant but technically complicated. It is not yet known what particles they describe or whether they are physically consistent.

The simplest supergravity theory describes a world consisting of gravitons and gravitinos only. That is clearly unrealistic since a unified field theory must have a place for all elementary particles. The number of such particles is not precisely known. It now appears that the quarks, which are the building blocks of protons and neutrons, are elementary. Other particles that must be included are the electron and three related particles: the muon and two kinds of neutrino. The various gauge particles must also be counted: the photon, the intermediate vector bosons of the weak interaction and the graviton. If supergravity is correct, the gravitino must be added to the list.

To accommodate these particles local supersymmetry must be extended to include states with a spin of less than 3/2. In principle this extension is easily achieved, since supersymmetry can connect any fermion-boson pair with adjacent spins. What is required, then, is that more particles be added to the theory in adjacent-spin doublets Techniques have been developed for describing the interactions of these doublets with the basic spin-2 and spin-3/2 doublet of the supergravity gauge fields. In the past year many such theories have been devised.

In one category of theories an arbitrary number of particles can be added in doublets of spin-1 and spin-1/2 or of spin-1/2 and spin-0. The advantage of these theories is that the number of particles can be adjusted almost at will to suit the elementary particles observed. A price must be paid, however, for this flexibility. Such theories have the undesirable characteristic that infinities in the one-loop diagrams cannot be eliminated. The reason is the lack of full unification. In these theories there are no symmetry transformations that connect the graviton and the gravitino to the lower-spin particles. Nevertheless, these theories may play an important role in the description of nature.

Another set of theories, known as extended supergravity theories, is much more restrictive and completely unified. There are just eight of these theories and each of them has a characteristic number of distinct boson-fermion transformations; this number is designated n, and it can assume values from 1 through 8. In each theory there is one spin-2 graviton and there are n spin-3/2 gravitinos. The number of particles with lower spins is also completely determined. If n is equal to 1, the theory is simply supergravity in its original form, with one graviton and one gravitino. If n is 2, the theory includes one graviton, two gravitinos and one spin-1 particle. Perhaps the most realistic model of this kind is given when n is 8. The complement of elementary particles then consists of one graviton, eight gravitinos, 28 spin-1 particles, 56 spin-1/2 particles and 70 spin-0 particles.

An intriguing property of the eight extended supergravity theories is their extreme degree of symmetry. Each particle is related to particles with adjacent values of spin by supersymmetry transformations, and these supersymmetries are of local form. Thus a graviton can be transformed into a gravitino and a gravitino into a spin-1 particle. Within each family of particles that have the same spin all the particles are related by a global internal symmetry, much like the isotopic symmetry that relates proton and neutron. Hence any gravitino can be transformed into any other gravitino by an internal symmetry operation. By combining the supersymmetry and the internal symmetry the entire group of particles is unified. One could start with the graviton and by a series of supersymmetry operations and internal-symmetry operations transform it into any other elementary particle that is included in the theory.

Like other physical symmetries, extended supergravity can also be viewed in terms of a "superparticle" with an arrow in an auxiliary space of many dimensions. As the arrow rotates, the particle becomes in turn a graviton, a gravitino, a photon, a quark and so on. The quanta of all the forces are present in the theory, and they are unified, or derived from a common source. This degree of unification has never before been achieved in a quantum field theory.

One remaining requirement for unification is to make the internal symmetries, which relate particles that have the same spin, local rather than global. Local internal symmetry is necessary so that forces such as electromagnetism can be incorporated. The requirement of local invariance can in fact be satisfied, as was demonstrated by one of us (Freedman) and Ashok Das of City College of the City University of New York and by E. S. Fradkin and M. A. Vasiliev of the Lebedev Institute. The resulting

THEORY	PARTICLE CONTENT				
	SPIN = 0	SPIN = 1/2	SPIN = 1	SPIN = 3/2	SPIN = 2
$N = 1$				1	1
$N = 2$			1	2	1
$N = 3$		1	3	3	1
$N = 4$	2	4	6	4	1
$N = 5$	10	11	10	5	1
$N = 6$	30	26	16	6	1
$N = 7$	70	56	28	7	1
$N = 8$	70	56	28	8	1

EXTENDED SUPERGRAVITY THEORIES incorporate not only gravitons and gravitinos but also elementary particles of lower spin, some of which might correspond to known particles such as the photon and the electron. There are eight such extended theories, each designated by a number, n, equal to the number of spin-3/2 gravitinos included in the theory. If any of these theories is a valid description of nature, the ensemble of particles it predicts must include all the elementary particles in the universe. In this way all the elementary states of matter would be unified, but a disturbing problem remains: even the largest of the extended theories does not seem to have room enough for all the known spin-1/2 and spin-1 particles.

theories potentially unify gravitation with the strong, the weak and the electromagnetic forces.

In the extended supergravity theories the strength of the gravitational force is determined by one parameter and the strengths of all the other forces are determined by another parameter. Ideally in a unified theory all the forces would be governed by a single universal constant. There does exist a third class of supergravity theories that have this feature. The simplest of them was found by, among others, Michio Kaku of City College and Paul Townsend of the State University of New York at Stony Brook. These theories are not based on Einstein's general theory of relativity but are supersymmetric generalizations of another theory of gravitation, discovered by Hermann Weyl in 1923. As in Einstein's theory, gravitation is described by the curvature of space-time, but there is an additional local symmetry that allows the scale by which length and time are measured to be chosen arbitrarily at each space-time point. The corresponding supersymmetric Weyl theories achieve a complete fusion of gravity with all the other forces, but for now the other two types of supergravity theories are likely to be more successful in applications to the real world.

The principle of local supersymmetry leads to an elegant unification of the basic forces, but elegance is not enough. The theories must pass a test that has been failed by all earlier theories of quantum gravity: the infinities that appear in calculations of interaction probabilities must be eliminated.

In all quantum field theories the diagrams in which virtual particles form closed loops describe the genuine quantum effects. The computation of the probabilities associated with any such diagram requires the summation of the virtual particles over all possible energies. These sums usually lead to infinities in the mathematical expressions for the diagrams. In some cases, such as quantum electrodynamics, the infinities are of a relatively harmless nature and can be removed by the renormalization procedure. In quantum gravity, however, the infinities are much worse and cannot be removed by renormalization. One hope remains for eliminating them. In some diagrams the infinity is positive and in others it is negative. It is only the sum of the diagrams that gives a physically observable probability, and it is possible that the infinities might miraculously cancel one another in the sum.

A simple argument explains why the infinities in quantum gravitation are worse than those in quantum electrodynamics. The electric force between charged particles is independent of the masses or energies of the particles and depends only on their charge, whereas

the gravitational force is proportional to the masses. Since in special relativity mass and energy are related by $E = mc^2$, it follows that the strength of the gravitational force increases when the energies of the virtual particles increase. Thus in the sum of the probabilities contributed by all possible energies of the virtual particles the higher energies give a larger contribution in the case of gravitation and lead to more serious infinities. Roughly speaking the infinities are of the form of the mathematical series $1 + 2 + 3 + \ldots$ in gravitation and of the milder form $1 + 1 + 1 + \ldots$ in electromagnetism.

Before the development of supergravity a careful analysis of the one-loop diagrams was undertaken in the quantized form of general relativity. The desired cancellation was found in diagrams that contain gravitons but no other particles. This result was obtained by Bryce S. DeWitt of the University of Texas and by 't Hooft and Veltman. The infinities cancel because of a special property of four-dimensional space-time; they would not cancel if our world had some other dimensionality.

Since the world is not made of gravitons alone, this finding would seem to be of limited practicality. All attempts to retain the finite result in theories with both gravitons and other kinds of particles have failed; when the lower-spin particles are introduced, the infinities spring up again.

It was therefore a dramatic moment when an explicit calculation of a physical process, a process known to be plagued by infinities in Einstein's theory, turned out to be finite in supergravity. Ensuing theoretical work and other explicit calculations have shown that in each of the eight theories of extended supergravity the sum of all diagrams with one loop is finite for all physical processes. This amelioration of the problem of infinities in quantum gravity is a most encouraging feature of supergravity.

The exact cancellation of some dozens of infinite terms seems too extraordinary to be entirely fortuitous, and it has a simple explanation. Consider first the subset of loop diagrams in which all the incoming and outgoing particles are gravitons; other particles can appear in these diagrams only as virtual particles. The sum of all the diagrams in this set can be shown to be finite by an extension of the similar result obtained in general relativity.

In the eight theories of extended supergravity finite results can also be obtained in diagrams with incoming and outgoing particles of lower spin. The infinities cancel in these diagrams because of the full unification of the theories. Diagrams with incoming photons or outgoing electrons, for example, can be related to diagrams that have only grav-

itons by applying the symmetry transformations available in these theories. In effect all the diagrams can be reduced to those that have only gravitons, and diagrams containing only gravitons are already known to have a finite sum.

The infinities cancel only in extended supergravity theories because only in those theories can all particles be transformed into gravitons. Other supergravity theories that include an arbitrary number of spin-0, spin-1/2 and spin-1 particles lack a crucial symmetry between the spin-1 particles and the spin-3/2 gravitinos. In the theories based on Weyl's theory of gravitation infinities present no problem in diagrams with any number of loops. On the other hand, in these theories it is not known whether probabilities always have positive values, which is a necessary criterion for any sensible physical theory.

The process in which the cancellation of infinities was first discovered by explicit calculation was the scattering of one photon by another photon. In general relativity each of the one-loop diagrams for this process has an associated probability amplitude that consists of an infinite quantity multiplied by a coefficient, which is generally a fraction. The sum of the coefficients is greater than zero and so the probability amplitude for the scattering is infinite. In supergravity additional diagrams that contain gravitinos must be added to the calculation. Each of these diagrams is also infinite, but the sum of the coefficients for all the diagrams is now zero.

Explicit calculations for other processes in supergravity have now been completed. Further theoretical arguments have been found indicating that in the eight extended supergravity theories the infinities also cancel in the sum of all two-loop diagrams. The finite results were first obtained by one of us (van Nieuwenhuizen) with Marcus T. Grisaru of Brandeis and J. A. M. Vermaseren of Purdue University.

Supergravity is a significant theoretical development because it offers hope of solving important, long-standing problems in physics: the unification of the fundamental forces and the elimination of infinities in quantum gravity. In the unification of fermions and bosons and in the derivation of all forces from the common requirement of local symmetry one can glimpse a deeply satisfying order in the theory. It remains to be determined, however, whether that order exists in nature. Several difficulties in the interpretation of the theory remain to be overcome.

One problem arises from the requirement that the internal symmetry in extended supergravity be a local one. In going from a global to a local symmetry an unexpected term is introduced into the equations: it is called the cosmological term, and it was first discussed by Einstein in early applications of general relativity to cosmology. The effect of the cosmological term is to assign a finite size to the universe. The emergence of the term in this context is puzzling: the internal symmetry has to do with the strength of electromagnetic and nuclear forces, and it seems curious that they should affect the size of the universe. What is worse, the value of the cosmological term predicted by the theory exceeds the upper limit derived from observational evidence.

Another conflict with observation is even more conspicuous. All particles in the extended supergravity theories are massless, but there is no question that many real particles, such as the electron, have a nonzero mass. A promising approach to the problem is to assume that some of the particles in extended supergravity acquire a mass through the mechanism of spontaneous symmetry-breaking. This might explain why the fundamental gravitinos predicted by extended supergravity have not been observed. Their mass might be so large that the accelerators available today have too little energy to create them. It is intriguing to note that spontaneous symmetry-breaking also changes the cosmological term in a quantum field theory. The question of whether it could reduce the term in extended supergravity is currently under study.

An important requirement if supergravity is to be considered a realistic theory is a definitive proof that the infinities cancel in the sum of all loop diagrams. So far cancellation has been demonstrated only for diagrams with one loop or two loops. Diagrams with three or more loops are no less important. Demonstrating that they are finite will probably require qualitatively different mathematical techniques. Most desirable would be a general proof that the sum of all possible diagrams is finite.

The specificity of extended supergravity must be counted among its virtues rather than its failings: the eight theories have few free or adjustable parameters and so they make well-defined predictions. Indeed, each of the theories provides a complete list of all the elementary particles in nature. The predictions, however, do not entirely correspond to the list of elementary particles known today. The most promising theory results when n is equal to 8. This theory is the largest one possible in extended supergravity, and some of its families of particles bear a tantalizing resemblance to groups recognized in nature. For example, Murray Gell-Mann of Cal Tech has recently shown that after spontaneous symmetry-breaking the theory correctly predicts certain properties of quarks, such as their electric charge. The prediction is all the more remarkable in that quarks, among all elementary particles, are assigned fractional electric charges. On the other hand, the same theory has a serious shortcoming. There are not enough places for other known particles such as the muon and the intermediate vector bosons.

These problems must ultimately be resolved by further study of supergravity theories and perhaps by their revision. It may develop instead, however, that what is in need of revision is current opinion as to which particles in nature are elementary.

BIBLIOGRAPHIES

I BACKGROUND: QUANTUM MECHANICS AND PARTICLE PHYSICS

1. What Is Matter?

SCIENCE AND HUMANISM: PHYSICS IN OUR TIME. Erwin Schrödinger. Cambridge University Press, 1951.

THE REVOLUTION IN PHYSICS. Louis de Broglie. The Noonday Press, 1953.

2. Field Theory

QUANTUM ELECTRODYNAMICS. F. J. Dyson in *Physics Today.* September, 1952.

THE STRUCTURE OF MATTER. Francis Owen Rice and Edward Teller. John Wiley & Sons, Inc., 1949.

3. Elementary Particles

CHARGE INDEPENDENCE THEORY OF V PARTICLES. Kazuhiko Nishijima in *Progress of Theoretical Physics,* Vol. 13, No. 3, pages 285–304; March, 1955.

ELEMENTARY PARTICLES. Enrico Fermi. Yale University Press, 1951.

THE INTERPRETATION OF THE NEW PARTICLES AS DISPLACED CHARGE MULTIPLETS. M. Gell-Mann in *Nuovo Cimento,* Supplemento al Vol. 4, Series 10, No. 2, 2° Semestre, pages 848–866; 1956.

4. Strongly Interacting Particles

DISPERSION AND ADSORPTION OF SOUND BY MOLECULAR PROCESSES. Proceedings of the International School of Physics, Italian Physical Society Course 27, 1962. Academic Press, 1964.

AN INTRODUCTION TO ELEMENTARY PARTICLES. W. S. C. Williams. Academic Press, 1961.

STRANGE PARTICLES. Robert Kemp Adair and Earle Cabell Fowler. Interscience Publishers, Inc., 1963.

TRACKING DOWN PARTICLES. R. D. Hill. W. A. Benjamin, Inc., 1963.

II CURRENT CONCEPTS: QUARKS AND QUANTUM CHROMODYNAMICS

5. Electron-Positron Annihilation and the New Particles

THE WORLD OF ELEMENTARY PARTICLES. Kenneth W. Ford. Xerox College Publishing, 1963.

PHYSICS IN THE TWENTIETH CENTURY: SELECTED ESSAYS. Victor F. Weisskopf. The MIT Press, 1972.

DISCOVERY OF A NARROW RESONANCE IN e^+e^- ANNIHILATION. M. L. Perl, B. Richter, W. Chinowsky, G. Goldhaber, G. H. Trilling et al. in *Physical Review Letters,* Vol. 33, No. 23, pages 1406–1408; December 2, 1974.

EXPERIMENTAL OBSERVATION OF A HEAVY PARTICLE *J*. Samuel C. C. Ting et al. in *Physical Review Letters,* Vol. 33, No. 23, pages 1404–1406; December 2, 1974.

DISCOVERY OF A SECOND NARROW RESONANCE IN e^+e^- ANNIHILATION. G. Goldhaber, G. H. Trilling, M. L. Perl, B. Richter et al. in *Physical Review Letters,* Vol. 33, No. 24, pages 1453–1455; December 9, 1974.

6. Quarks with Color and Flavor

WEAK INTERACTIONS WITH LEPTON-HADRON SYMMETRY. S. L. Glashow, J. Iliopoulos and L. Maiani in *Physical Review D*, Vol. 2, No. 7, pages 1285–1292; October 1, 1970.

FACT AND FANCY IN NEUTRINO PHYSICS. A. De Rújula, Howard Georgi, S. L. Glashow and Helen R. Quinn in *Reviews of Modern Physics*, Vol. 46, No. 2, pages 391–407; April, 1974.

RECENT PROGRESS IN GAUGE THEORIES OF THE WEAK, ELECTROMAGNETIC AND STRONG INTERACTIONS. Steven Weinberg in *Reviews of Modern Physics*, Vol. 46, No. 2, pages 255–277; April, 1974.

HADRON MASSES IN A GAUGE THEORY. A. De Rújula, H. Georgi and S. L. Glashow in *Physical Review D*, Vol. 12, No. 1, pages 147–162; July 1, 1975.

7. The Confinement of Quarks

NEW EXTENDED MODEL OF HADRONS. A. Chodos, R. L. Jaffe, K. Johnson, C. B. Thorn and V. F. Weisskopf in *Physical Review D*, Vol. 9, No. 12, pages 3471–3495; June 15, 1974.

ASYMPTOTIC FREEDOM: AN APPROACH TO STRONG INTERACTIONS. H. David Politzer in *Physics Reports*, Vol. 14C, No. 4, pages 129–180; November, 1974.

DUAL-RESONANCE MODELS OF ELEMENTARY PARTICLES.
John H. Schwarz in *Scientific American*, Vol. 232, No. 2, pages 61–67; February, 1975.

MASSES AND OTHER PARAMETERS OF THE LIGHT HADRONS. T. DeGrand, R. L. Jaffe, K. Johnson and J. Kiskis in *Physical Review D*, Vol. 12, No. 7, pages 2060–2076; October 1, 1975.

EXTENDED SYSTEMS IN FIELD THEORY: PROCEEDINGS OF THE MEETING HELD AT ÉCOLE NORMALE SUPÉRIEURE, PARIS, JUNE 16–21, 1975 in *Physics Reports*, Vol. 23C, No. 3, pages 237–374; February, 1976.

8. The Bag Model of Quark Confinement

MASSES AND OTHER PARAMETERS OF THE LIGHT HADRONS. T. DeGrand, R. L. Jaffe, K. Johnson and J. Kiskis in *Physical Review D*, Vol. 12, No. 7, pages 2060–2076; October 1, 1975.

STRINGLIKE SOLUTIONS OF THE BAG MODEL. K. Johnson and C. B. Thorn in *Physical Review D*, Vol. 13, No. 7, pages 1934–1939; April 1, 1976.

THE CONFINEMENT OF QUARKS. Yoichiro Nambu in *Scientific American*, Vol. 235, No. 5, pages 48–60; November, 1976.

QUARK CONFINEMENT. R. L. Jaffe in *Nature*, Vol. 268, No. 5617, pages 201–208; July 21, 1977.

III FUTURE DIRECTIONS: THE SEARCH FOR A UNIFIED THEORY

9. Unified Theories of Elementary-Particle Interaction

RELATIVISTIC QUANTUM FIELDS. J. D. Bjorken and S. D. Drell. McGraw-Hill Book Co., Inc., 1965.

A MODEL OF LEPTONS. S. Weinberg in *Physical Review Letters*, Vol. 19, No. 21, pages 1264–1266; November 20, 1967.

THEORY OF WEAK INTERACTIONS IN PARTICLE PHYSICS. R. E. Marshak, Riazuddin and C. P. Ryan. Wiley-Interscience, 1969.

GAUGE THEORIES. E. S. Abers and B. W. Lee in *Physics Reports*, Vol. 9C, No. 1. North Holland Publishing Company, 1973.

SPONTANEOUS SYMMETRY BREAKING, GAUGE THEORIES, THE HIGGS MECHANISM AND ALL THAT. Jeremy Bernstein in *Reviews of Modern Physics*, Vol. 46, pages 7–48; January, 1974.

RECENT PROGRESS IN GAUGE THEORIES OF THE WEAK, ELECTROMAGNETIC AND STRONG INTERACTIONS.
Steven Weinberg in *Reviews of Modern Physics*, Vol. 46, No. 2, pages 255–277; April, 1974.

10. Supergravity and the Unification of the Laws of Physics

CONSISTENT SUPERGRAVITY. Stanley Deser and Bruno Zumino in *Physics Letters*, Vol. 62B, No. 3, pages 335–337; June 7, 1976.

PROGRESS TOWARD A THEORY OF SUPERGRAVITY. Daniel Z. Freedman, Peter van Nieuwenhuizen and Sergio Ferrara in *Physical Review D: Particles and Fields*, Vol. 13, No. 12, pages 3214–3218; June 15, 1976.

ONE-LOOP RENORMALIZABILITY OF PURE SUPERGRAVITY AND OF MAXWELL-EINSTEIN THEORY IN EXTENDED SUPERGRAVITY. Marcus T. Grisaru, Peter van Nieuwenhuizen and J. A. M. Vermaseren in *Physical Review Letters*, Vol. 37, No. 25, pages 1662–1666; December 20, 1976.

INDEX